Low-Temperature Systems

Edited by Tatiana Morosuk

Published in London, United Kingdom

Low-Temperature Systems
http://dx.doi.org/10.5772/intechopen.1004481
Edited by Tatiana Morosuk

Contributors
Abhay Singh Gour, Alejandro Zacarías, Ankit Anand, Chaohui Liu, Cuauhtémoc Jiménez, Guerlin Romage, Haidong Yan, Ignacio Carvajal, Ivoni Carlos Acunha Jr., José de Jesús Rubio, María Venegas, Tripti Sekhar Datta, Vutukuru Vasudeva Rao, Wangjun Cheng, Wanli Li, Yaqin Liao, Yuriy Lobunets

First published in London, United Kingdom, 2025 by IntechOpen
IntechOpen is the global imprint of INTECHOPEN LIMITED, registered in England and Wales, registration number: 11086078, 167-169 Great Portland Street, London, W1W 5PF, United Kingdom

For EU product safety concerns: IN TECH d.o.o., Prolaz Marije Krucifikse Kozulić 3, 51000 Rijeka, Croatia, info@intechopen.com or visit our website at intechopen.com.

British Library Cataloguing-in-Publication Data
A catalogue record for this book is available from the British Library

Low-Temperature Systems
Edited by Tatiana Morosuk
p. cm.
Print ISBN 978-1-83634-101-7
Online ISBN 978-1-83634-100-0
eBook (PDF) ISBN 978-1-83634-102-4

Meet the editor

Professor Tatiana Morosuk is the Head of the department for Exergy-Based Methods for Refrigeration Systems at Technische Universität Berlin in Germany. She received her Diploma, Ph.D., and Doctor Habilitatus degree, as well as a professorship, all in Ukraine. She supervised over 20 Ph.D. and 100 M. Sc. theses, published eight books, and authored over 400 research papers. She is the study dean of the International Master's program "Process, Energy, Environmental System Engineering". Tatiana Morosuk is the first female recipient of the ASME Potter Gold Medal (American Society of Mechanical Engineers) for outstanding and innovative contributions to theoretical and applied thermodynamics, teaching, and research in advanced exergy-based methods, refrigeration, cryogenic processes, and power generation plants. She is the Editor-in-Chief of the Journal of Energy Resources Technology (ASME).

Contents

Preface

Energy engineering and applied physics are rapidly evolving in response to growing demands for higher energy efficiency, reduced environmental impact, and flexible operation systems. These demands are further intensified by global trends, such as e-mobility and digitalization, with the ultimate goal of transitioning to carbon-neutral energy systems. A key contributor to this transformation is the application of low-temperature technologies (cryogenics and superconductivity), which are now increasingly vital in a broad range of fields, including power electronics, advanced energy storage, precision materials processing, semiconductor packaging, and thermal management system design. Leveraging the unique thermal, electrical, and mechanical properties that materials exhibit at reduced temperatures enables engineers to enhance device reliability, operational efficiency, and system-level integration, thereby paving the way for innovation across various sectors, from aerospace to renewable energy.

The successful IntechOpen book project, "Low-temperature Technologies" [1], provided a comprehensive overview of refrigeration technologies, with a primary focus on basic principles and established systems. A new edited book, in contrast, intensely focuses on five interrelated, cutting-edge areas of low-temperature applications for industrial refrigeration systems, multi-stage thermoelectric heat pumps, power electronics, superconducting energy storage, and cryogenic metallurgy.

While each chapter addresses a different research and application field, they are unified by a shared dependence on low-temperature phenomena.

Chapter 1 - *Energy Efficiency in Industrial Refrigeration Systems* – evaluates energy optimization strategies for industrial refrigeration systems. These systems are among the most energy-intensive in industry and are major contributors to operating costs and carbon emissions. In addition to system modeling and thermodynamic analysis, real-world case studies are included. These examples illustrate how technical improvements can significantly reduce electricity consumption, extend system lifespan, and align operations with sustainability objectives.

Chapter 2 - *System Analysis of Thermoelectric Heat Pump Circuits with Regenerative Heat Recovery* – introduces the novel design features of solid-state thermoelectric heat pumps. These systems operate using thermoelectric modules arranged in a multi-stage configuration. Each module handles a small temperature gradient, allowing the system to maintain high thermal conversion efficiency. This approach also enables regenerative heat recovery, significantly improving overall system performance.

Chapter 3 - *Adiabatic and No-Adiabatic Absorption Chillers Using Ammonia-Lithium Nitrate Solutions* – reports the detailed modeling and analysis of this kind of thermally-driven refrigeration systems.

Chapter 4 - *Low-Temperature Sintering Technologies in Power Electronics: Materials, Process, and Advanced Packaging of SiC WBG Semiconductors* – focuses on superconducting magnetic energy storage systems. These systems store electrical energy in the magnetic field generated by a persistent current flowing through a superconducting coil, thereby eliminating resistive losses and enabling fast charge and discharge cycles. These systems are particularly effective for grid stabilization, improving power quality, and supporting critical and peak loads.

Chapter 5 - *Energy Storage with Superconducting Magnets: Low-Temperature Applications* – dedicated to low temperatures used in the packaging of wide-bandgap semiconductor devices, such as silicon carbide and gallium nitride. These materials support high-voltage, high-frequency, and high-temperature operation, making them critical for next-generation power systems.

Chapter 6 - *Low-Temperature Performance and Manufacture of Metals* – investigates the mechanical behavior and processing of metals at low and cryogenic temperatures. Materials like aluminum alloys, copper, and stainless steel exhibit changes in yield strength, ductility, and strain-hardening behavior under these conditions. Here, the results from testing, fracture analysis, and microstructural evaluation explain how temperature affects deformation mechanisms at both macro- and micro-scales.

This book seeks not only to inform but to inspire further cross-disciplinary research and practical deployment of low-temperature technologies. This book is particularly important for master's and PhD students, as well as researchers.

Tatiana Morosuk
Professor,
Technische Universität Berlin,
Berlin, Germany

Reference

[1] Morosuk T, Sultan M, editors. Low-temperature Technologies. InTech Open; 2019

Chapter 1

Energy Efficiency in Industrial Refrigeration Systems

Ivoni Carlos Acunha Jr.

Abstract

Energy efficiency in industrial refrigeration systems should be an object of study, especially large ones used for producing and storing food and beverage products. This is because this system requires large electricity consumption and, consequently, carries out environmental impacts. Some strategies and technologies can be used to increase the coefficient of performance (COP) of refrigeration units, such as intelligent operation through variable speed drives (VSDs) in pumps and fans, floating head pressure work, optimization of ice and chilled water production, intelligent controls in condensers and compressors, use of mathematical modeling and computer simulations, among others. Therefore, this work aims to highlight the impact of strategy and technology employment on energy efficiency improvements at industrial refrigeration systems by bringing studies of cases about refrigeration units used in food refrigeration of dairy, pork, and poultry products.

Keywords: industrial refrigeration systems, food refrigeration, energy efficiency, computational simulation, variable speed driver

1. Introduction

The functionality and construction costs are frequently given as priority in some industrial refrigeration systems. The energy costs of a refrigeration plant in cooling and freezing foods and beverages can achieve 75% of the overall electrical consumption in industries of foods. Energy saving can be possible by employing operational strategies.

In this work are presented some results of cases analyzed by computational simulation and equipment measurements in some food industries. The computational simulations are done using the SGEM® system, which was built to find optimal operational parameters of the compressors, condensers, evaporators, ice machines, water coolers, poultry chillers, and others operating together in an industrial refrigeration system. The compressor is the component of the refrigeration system that has the biggest demand for electricity, to the point of significantly affecting the operating cost of the installation. Therefore, it is very important to have the correct definition of compressors' operating pressures, and proper component selection during the design system, to attend to operating conditions and ensure the best energy efficiency [1].

Eikevik [2] realized an experimental analysis of five screw compressors with ammonia as the refrigerant fluid, developing an optimal model for compression operation to increase energy efficiency. Hovgaard et al. [3] describe a novel economic-optimizing Model Predictive Control scheme to reduce operating costs by utilizing the thermal storage capabilities. A nonlinear optimization tool to handle a non-convex cost function is utilized for simulations. They formulate a new cost function that enables the refrigeration system to the balancing power market. That work discusses a novel incorporation of probabilistic constraints and Second Order Cone Programming with economics. Your formulations demonstrate how it can reduce the operational costs of the system. Plessis et al. [4] and Peng and Du [5] presented research with different techniques to control the capacity of heat exchangers and compressors and the use of variable speed drives (VSDs), highlighting the reduction in electricity consumption compared to standard installation or other control techniques. This study aims to present results from the use of computational simulations to predict behaviors of industrial refrigeration systems for the definition and use of operational strategies that resulted in increased energy efficiency in industrial refrigeration systems. Oh et al. [6] studied the common thinking that equipment should always be extensively reconfigured when switching from pure to mixed refrigerants. To determine the most energy-efficient operating conditions for each refrigeration design, an optimization framework is utilized, linking a process simulator with an external optimization method. For the case considered in this paper, savings of shaft power required for the refrigeration cycle can be achieved from 16.3 to 27.2% when the pure refrigerant is replaced with mixed refrigerants and operating conditions are re-optimized. Yin et al. [7] presented a novel energy-efficiency-oriented cascade control strategy for refrigeration systems to improve energy efficiency and fulfill the cooling requirements of indoor occupants simultaneously. A mathematical model is developed to determine the set point of superheat by a PI controller based on the nonlinear correlation between cooling demands and superheat degree. The pressure difference and superheat degree of the evaporator were controlled by a model predictive control strategy. That work demonstrates that it can improve energy efficiency by up to 5.8% when compared to another scheme. Cirera et al. [8] proposed a data-driven methodology that improves the efficiency of the refrigeration systems acting on the load side. The solution approaches the problem with a novel load management methodology that considers the estimation of the individual load consumption and the necessary robustness to be applicable in highly variable industrial environments. This study demonstrated the ability to reduce the electrical consumption of the compressors by 17% as well as a 77% reduction in the operation time of two compressors working in parallel, a fact that enlarges the machine's life. In this sense, this work intends to present some methodologies, strategies, and results guided by analyses based on mathematical modeling and computational simulations to improve the energy efficiency of industrial refrigeration systems. The main difference of this work refers to the methodology adopted, which is not used in industrial refrigeration systems, especially those that use ammonia as a refrigerant. The methodology and analyses employed are not verified in large refrigeration systems, given the complexity of analyzing these systems. All analyses presented here refer to case studies of industrial systems.

2. Analysis and strategies to COP improvement

For many years, industrial refrigeration systems have been used for different purposes. However, some technological advances currently implemented in smaller systems do not apply to large refrigeration units due to their specificities. More detailed analyses of large systems require a lot of time and large investments given the complexity of such systems. A thorough assessment can be made through mathematical modeling followed by computer simulations dedicated to each system since it is difficult to find identical systems for similar purposes on the market. One of the biggest challenges of this type of analysis is the field validation of the computer models developed since the equipment used no longer has the performance defined by the manufacturers due to issues ranging from correct application to efficiency losses due to continuous use over the years.

Thus, some parameters, designs, strategies, and technologies that can be used to increase the coefficient of performance (COP) of refrigeration units will be addressed below.

2.1 Energy assessment of different refrigeration systems

2.1.1 Refrigeration system for freezing poultry

Correct assessment of refrigeration facilities, as well as the sizing of their equipment, can make all the difference in the use of cold adopted. Among the main items, we can mention: The correct assessment of the thermal load of environments and processes; determination of the capacity of compressors for the regime used; the capacity of evaporators and condensers; the capacity of liquid pumps, liquid separators, valves, and other accessories.

By comparing the thermal load of a regime used for freezing and storing chickens and the refrigeration capacity of the compressors used for this purpose in a refrigeration unit evaluated, the graph in **Figure 1** is constructed.

From the graph, it is possible to observe that the thermal capacity of compressors available is much higher than the capacity required by the thermal load imposed. This excess capacity results in a substantial increase in energy to the system. Observing the energy waste of the compressors every hour, we arrive at the graph in **Figure 2**.

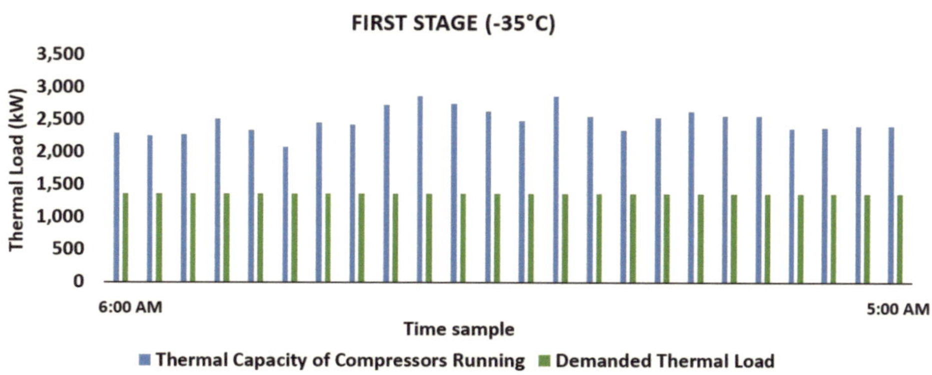

Figure 1.
Actual required thermal load versus thermal capacity of compressors in operation.

Through this graph, it is possible to observe the waste of up to 51% of electrical energy in the use of compressors in the machine room. At no time was the power waste in the compressors less than 40%.

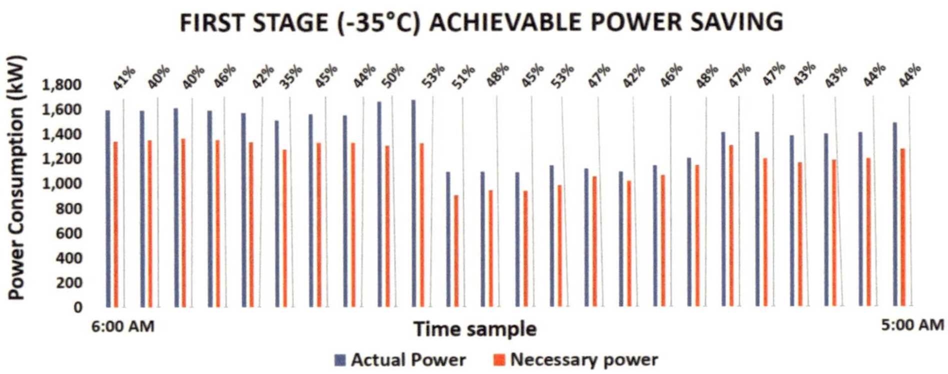

Figure 2.
Actual required electrical power versus required power of compressors in operation.

Therefore, the correct assessment of refrigeration installations and the correct dimensioning of their equipment can make all the difference in the use of cold adopted. **Figures 1** and **2** clearly show discrepancies regarding the correct dimensioning of thermal loads and the use of necessary cold.

The important task of accurately verifying the thermal balances between the heat transfer rate dissipated in the condensers and the total thermal capacity of the installed condensers is now highlighted. In this specific case, the thermal balances and their discrepancies are presented.

By comparing the heat transfer rate that must be dissipated in the condensers and the total heat rejection capacity of the installed condensers, the graph in **Figure 3** is constructed.

From the graph, it is possible to see that the condensation pressure is always above what should be expected, reaching a maximum value of 3.0 bar at 2:00 pm. This means that the unit's refrigeration compressors are significantly overconsuming electrical energy and losing refrigeration thermal capacity.

There are many reasons for this to occur, but the main ones may be linked to: the quality of the sprinkler water, distribution of sprinkler water, volume of sprinkler water, condensers with encrustation, water pumps with low flow rate, presence of

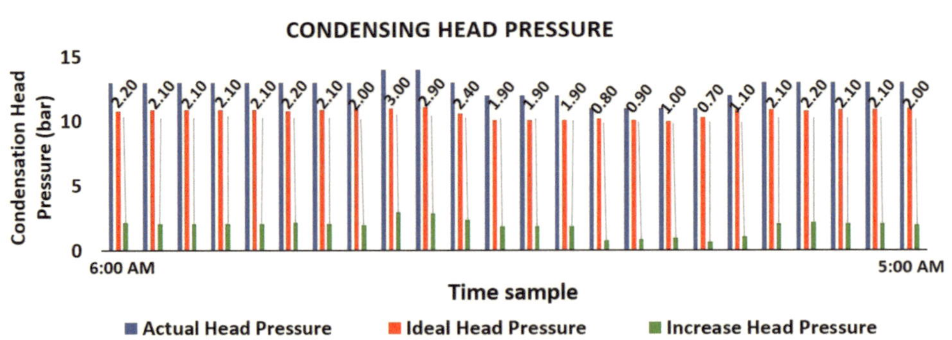

Figure 3.
Actual condensing head pressure versus possible condensation head pressure.

non-condensable, fans out of operation, shutdown of condensers by the staff at inappropriate times, even under low discharge pressure, and other causes of lesser impact.

The graph in **Figure 4** shows the possibility of energy saving each hour that the system operates.

As can be seen, the energy waste is considerable, reaching its maximum at 2:00 p.m., reaching 21.7%. At no time was it possible to verify the operation without wasting energy, and at 8:00 p.m., the lowest waste was 5.8%.

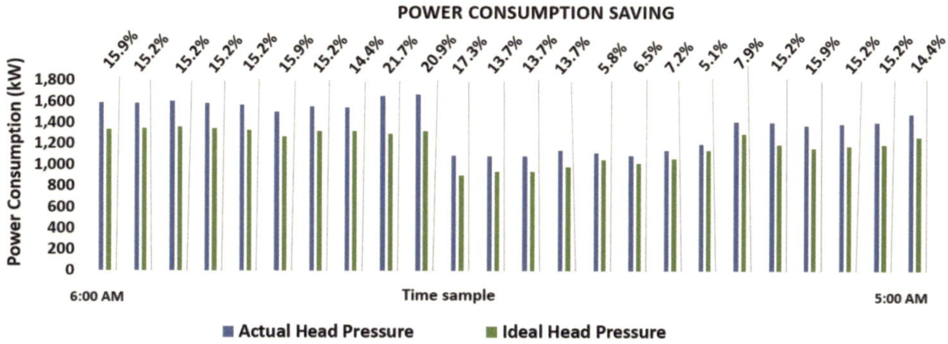

Figure 4.
Comparative power operating.

The total power that can be saved in this system, taking into account losses due to discharge overpressure and inadequate balances in tunnels and chambers, is shown in the graph in **Figure 5**.

At 3:00 p.m., the refrigeration system presents the best opportunities, reaching a possible saving in the power demand of 1000 kW. Although possibilities for lower gains can be verified, the smallest saving that can be achieved is approximately 527 kW, at 9:00 p.m. Adding up all the possibilities of gain presented in this evaluation, the annual saving that this system can present is 4,473,467.57 kWh over 1 year of operation, considering 22 days of operation per month.

Often, difficult access to condensers, reduced work teams, and lack of information on equipment operation in real-time, in addition to other aggravating factors, result in poor operating conditions that are only verified after days and even months of energy waste. This makes dynamic and focused monitoring necessary. To prevent these and other energy wastes, it is recommended to implement an intelligent parameter management and appropriate remote monitoring system that can recommend the correct actions in real-time.

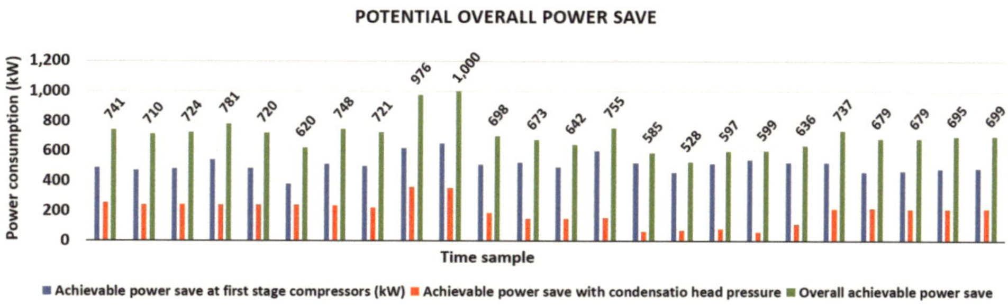

Figure 5.
Realizable power saving.

2.2 Refrigeration system for dairy products

Chilled water storage systems can be used for different purposes. One example is its use as an auxiliary reservoir for faster filling of chillers after sanitation operations, especially in refrigeration units that operate 24 h a day. Another strategy is storage for use during peak hours, thus reducing electricity demand at these times. Although many companies use the free energy market, some have high costs for distribution and use the electrical system during this period, significantly increasing the value of the electricity bill.

Figure 6 illustrates a chilled water-cooling system in which the system stores thermal energy during off-peak time for use during peak time. In this system configuration, compressors can be sized to be out of operation during peak hours, reducing electricity costs.

During off-peak times, both pumps are in operation, one for circulating chilled water in the heat exchangers directly involved in the process and the other for circulating water in the exchangers responsible for lowering the temperature of both the water circulating through the process exchangers and the water circulating in the thermal storage tank.

An important observation is that storage will also increase the thermal load of the system since the lateral surface of the tanks is generally large and exposed to the external environment. Although the thermal insulation of the tanks reduces the heating of the water, it will not completely prevent it, and an increase in compressor capacity will be allocated for this purpose. The performance of a chilled water production system is directly linked to several factors such as the performance of compressors, condensers, and ammonia quality, among others.

This analysis deals with a chilled water system with plate heat exchangers for immediate use and a storage tank to meet fluctuations in thermal load. **Figure 7** shows the flow rate of chilled water required at each hour of the day.

No less important is the assessment of the temperature of the chilled water returning from the production sector. Observing three different production sectors

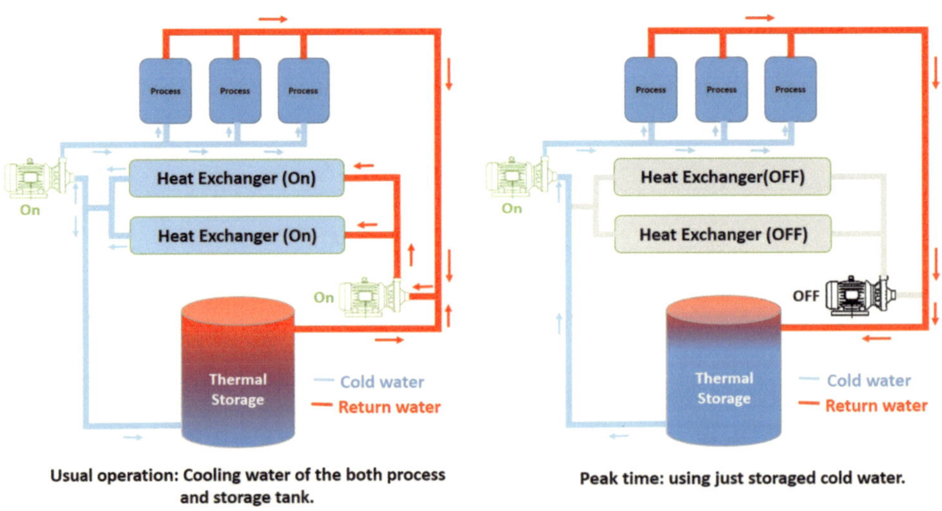

Figure 6.
Cold water system.

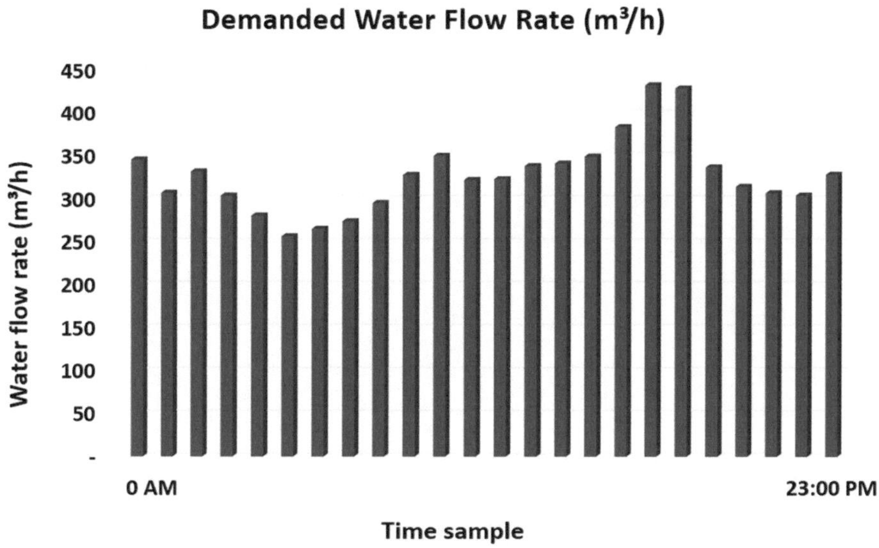

Figure 7.
Demanded water flow rate.

(milk reception, dairy processing, and serum processing), it can be seen (through **Figure 8**) that the temperatures can reach values that can reach 8°C in the dairy product return, 14°C in the whey return, and 10°C in the reception return, evidencing insufficient temporary capacity of the refrigeration system used, not allowing the maintenance of controlled temperatures within lower standards. In this case, better use of the chilled water storage system provides gains not only in electricity consumption but also in the quality of the cold produced. **Figure 8** shows the return temperatures of the following processes: (a) Milk reception, (b) dairy processing, and (c) serum processing.

The operational trend of capacity and consumption of the refrigeration system in the initially verified condition is shown in **Figure 9** for the conditions observed.

The blue curve represents the system consumption as a function of the ambient wet bulb temperature, while the red curve represents the discharge pressure. The curves refer to a water supply capacity of 760 m^3/h at 1.0°C with return at 4.0°C. For the condition evaluated, the estimated consumption of electrical energy is approximately 0.96 kWh for each cubic meter of water produced at 1°C, while in summer, this value will reach 1.03 kWh/m^3 of water supplied since the discharge pressure will increase from 11 bar to approximately 12.7 bar for the observed T_{WB} of 25°C. The implementation of strategies has brought possibilities for better operational standards, allowing the system to operate under the curve shown in **Figure 10**. This scenario can increase the capacity to supply chilled water at 1°C to 1200 m^3/h, ensuring a return to 4.0°C, representing an increase of approximately 58% compared to the current situation. Regarding the consumption of electrical energy required for production, this will increase to 0.90 kWh/m^3 of water produced in winter and 0.96 kWh/m^3 of water produced in summer, which can provide annual savings of approximately 648,000 kWh for the expected water consumption of 1200 m^3/h. The restoration of the system provided an increase in the availability of flow rate and a reduction in the cost of producing chilled water. In this specific case, the integration between the instantaneous production of melted water and tank storage proved to be the best solution. Another aspect to be considered is the discharge pressure, which directly impacts the consumption of refrigeration compressors.

The analysis of industrial units using modern technologies allows us to delve deeper into the identification of opportunities and significant energy savings.

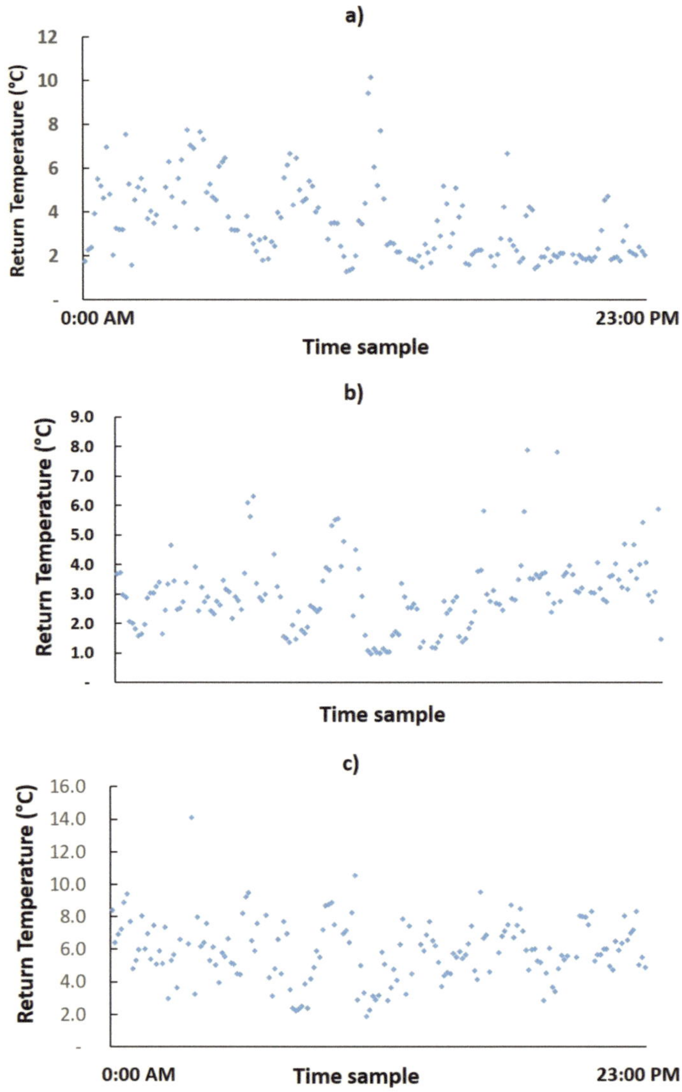

Figure 8.
Water return temperature – (a) Milk reception, (b) lacteous processing, and (c) serum processing.

2.2.1 Refrigeration system for pork products

At two different times, analyses were carried out to improve the performance of a refrigeration unit used for pig production, one in 2020 and the other in 2024. This refrigeration unit has 24 pig cooling chambers and two freezing tunnels with operating regimes in the machine room equal to –35/–10°C and – 10/+35°C. It is noteworthy

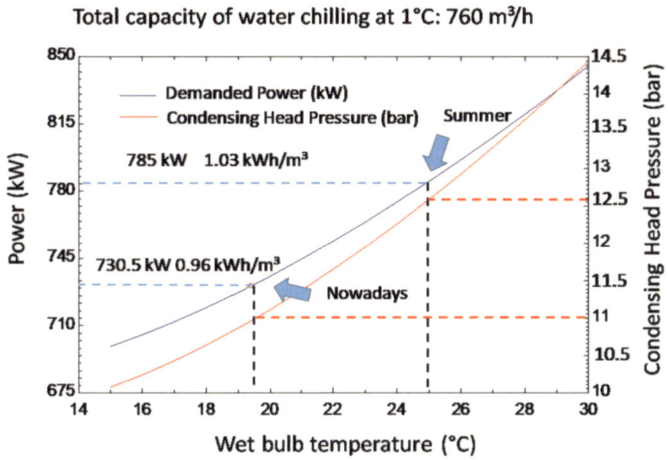

Figure 9.
Conditions of cold water production verified.

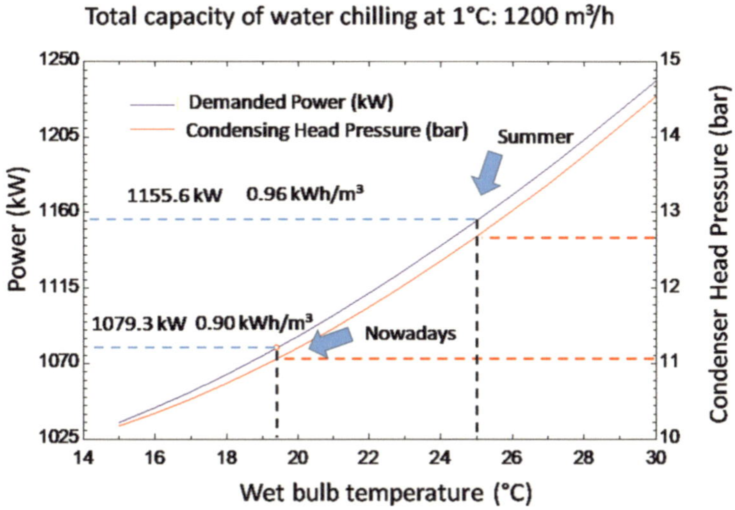

Figure 10.
Conditions of cold water production predicted after employment of strategies.

that units producing large animals such as meat and pork generally have a large number of cooling chambers with also large thermal amplitudes since the animals begin the cooling process at temperatures between 40 and 50°C and should finish at around 5°C after a period of 20 to 24 h. **Figure 11** shows the temperature variation in four different environments over 1 day of operation in a pork slaughterhouse with the respective expected setpoints: (a) palletizing room; (b) cooling chamber; (c) carton freezer tunnel; and (d) storage chamber.

These large temperature ranges sometimes cause difficulties in implementing setpoints that can meet thermal demands efficiently. **Figure 12** shows an example of a day of operation for a compressor connected to the low-temperature regime (−35°C). As can be seen, on that day the setpoint was not respected, and the lowest pressure

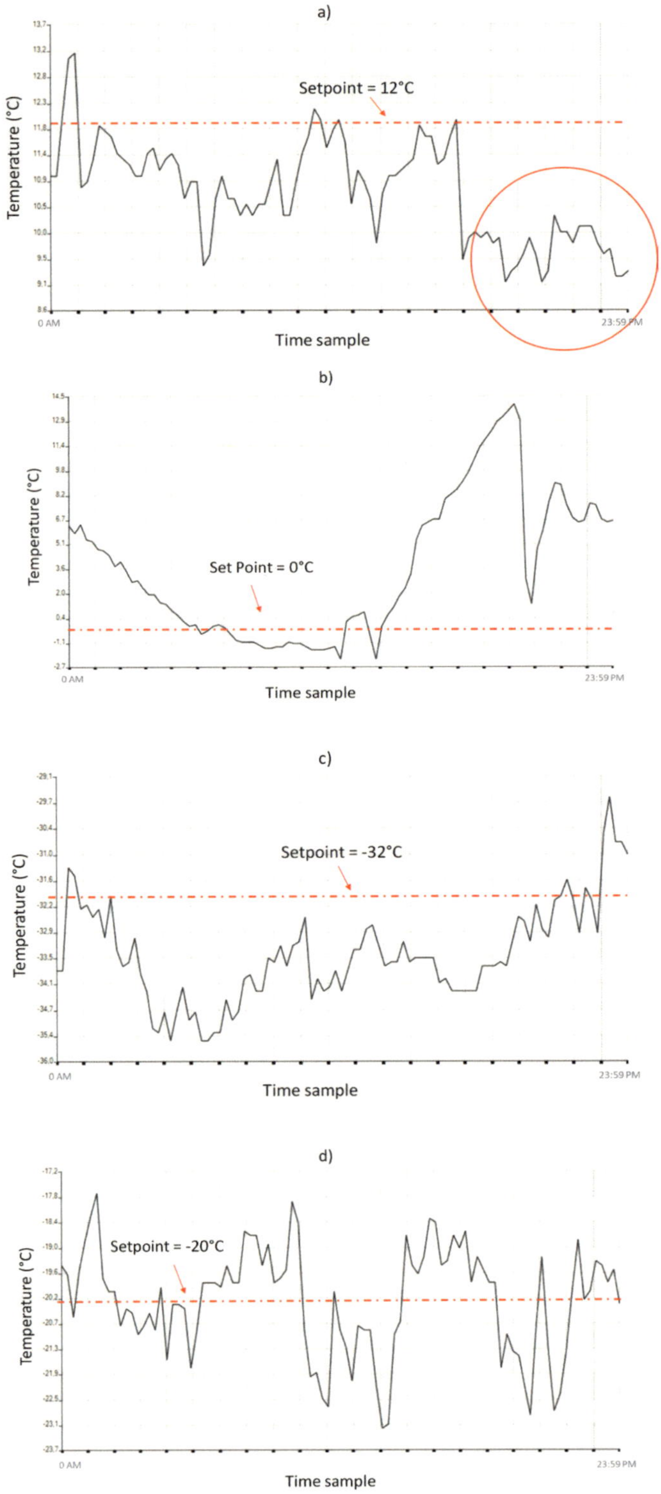

Figure 11.
Temperatures – (a) palletizing room; (b) cooling chamber; (c) carton freezer tunnel; and (d) storage chamber.

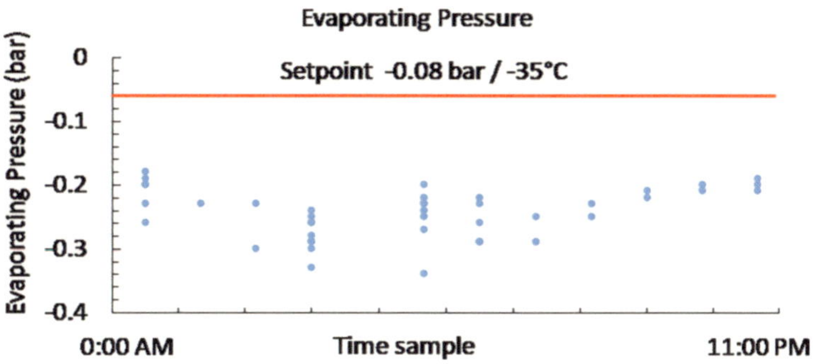

Figure 12.
Evaporating pressure.

recorded was −0.34 bar, which resulted in an energy waste of approximately 19.3% in the low-temperature regime.

Figure 13 shows the variation in intermediate pressure over one day of operation. The lowest intermediate pressure experienced on that day was 1.6 bar, which represents a waste of 10.3% of energy in the compressors connected to intermediate pressure. These fluctuations cause constant waste of energy, leading to major losses.

Computer simulations indicated strategies for peak and off-peak time. One of the strategies analyzed was the possibility of implementing the 0/+35°C regime by changing the suction pressure of the compressors instead of using pressure-regulating valves for this purpose. The pressure of the −35°C regime was changed from −0.4 bar to −0.1 bar, while the pressure of the intermediate regime went from 1.8 bar to 2.2 bar.

These strategies resulted in the shutdown of two and, at times, three compressors during off-peak time and four compressors during peak hours.

The annual savings generated were approximately 528,000 kWh during peak hours and 1,584,000 kWh during off-peak hours, representing an improvement of approximately 5.18 kWh per ton of processed product.

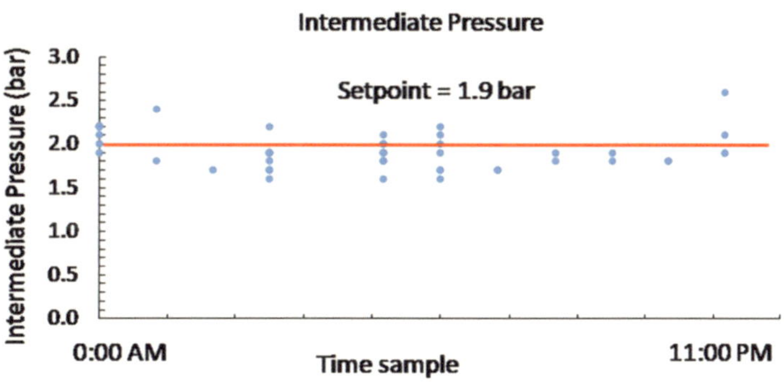

Figure 13.
Intermediate pressure.

2.3 Energy performance of evaporative condensers

As previously mentioned, condenser performance directly impacts the energy consumption of a vapor compression refrigeration system. Unfortunately, many of the condensers installed in the field have performances that are much lower than those predicted by their manufacturers. A detailed analysis of the operating conditions of evaporative condensers is shown below. Five evaporative condensers were analyzed in a pork slaughterhouse with five reciprocating compressors, four screw compressors, three carton freezers, one storage chamber, one chiller water with capacity of 120 m^3/h, and other environments necessary for pork processing. The head pressure condensing during the assessment was 10.4 bar. The first variable analyzed is the wet bulb temperature (T_{WB}) at the airflow inlet to the condenser (**Figure 14**).

The T_{WB} at the condenser air inlet directly impacts the energy gain of the air as it passes through the tube bundle. When observing the T_{WB} at the air inlet, it can be seen that condensers 2 and 5 should have better performance than the others, with condenser 1 having the worst performance.

Another important variable is the relative humidity at the condenser air outlet (**Figure 15**). This variable will indicate how efficient the air humidification is as it passes through the tube bundle. A desirable value is between 95 and 100%. The only condenser that presented adequate air saturation.

The energy gains of the air stream (**Figure 16**) are more intense in condensers 1, 2, and 4, evidencing poor use of the air as it passes through the tube bundle. Thus, despite low humidification, condensers 1 and 2 show good use of the air stream in terms of energy gain. A desirable use for evaporative condensers is around 20 kJ/kg in the air stream. Therefore, attention should be paid to the water and air flows in condensers 3 and 5.

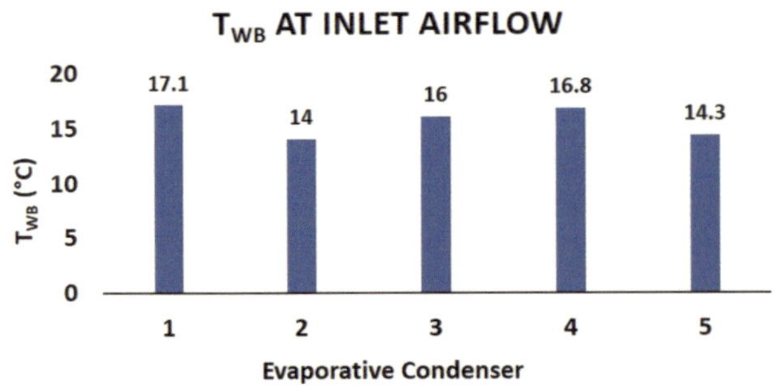

Figure 14.
Wet inlet bulb temperatures.

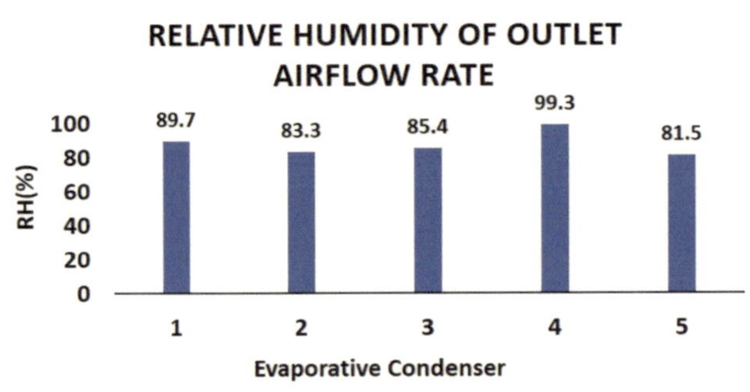

Figure 15.
Relative humidity of air flow rate.

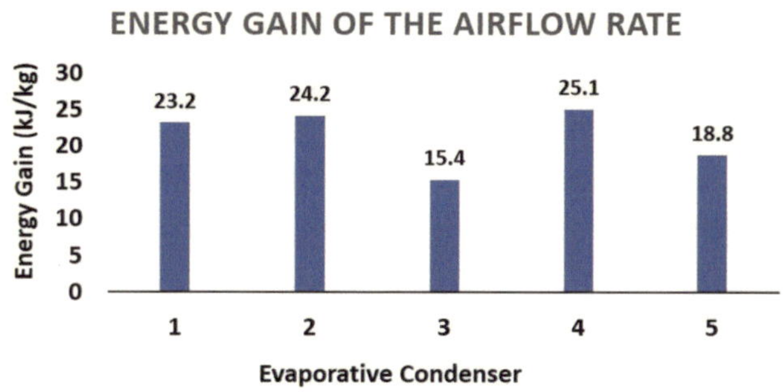

Figure 16.
Energy gain.

The heat transfer rate predicted by the manufacturers and the measured heat transfer rate for each of the ECs evaluated are shown in **Figure 17**. ECs 1 and 3 presented heat transfer rates much lower than those predicted by their manufacturers for the conditions analyzed. Condenser 2 presented a heat transfer rate higher than that predicted by the manufacturer. The coefficient of performance of the ECs is shown in **Figure 18**, evidencing the relationship between the heat transfer rate and the power consumed in each EC. Once again, it is possible to observe the low performance of ECs 1 and 3 and the good performance of EC 2. The best results were found in ECs 2, 4, and 5, and ECs 1 and 3 need to be revised to restore their standard operating conditions. The overall efficiency of ECs was 74%.

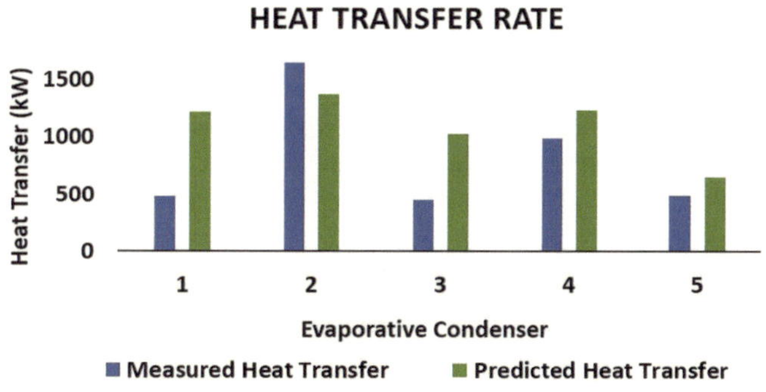

Figure 17.
Heat transfer rate.

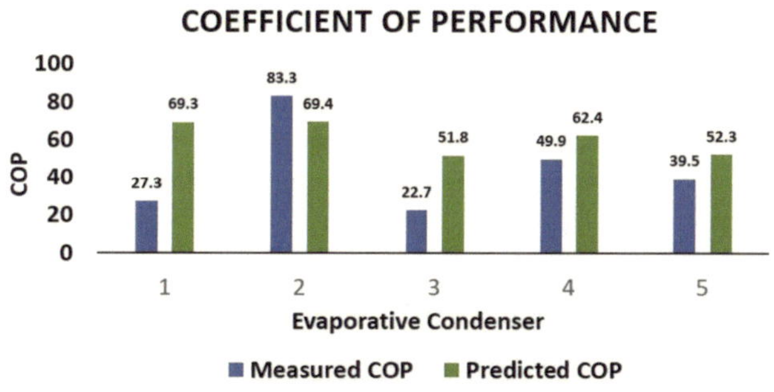

Figure 18.
Coefficient of Performance.

With the condensers operating at 100% of their predicted capacity, the condensing pressure should be 8.9 bar, which would lead to an energy saving of approximately 9.4% in the compressors, equivalent to an annual energy saving of 998,784 kWh.

3. Energy saving employing float head pressure controls

Compressor operating pressures play a fundamental role in both the capacity and mechanical power required to compress the refrigerant fluid. Suction pressure often corresponds to the temperature at which the fluid is pumped to the evaporators, directly influencing the capacity of this equipment and, consequently, the temperature of environments and products. Discharge pressure can be observed basically for two types of operation:

1-stage compression systems in which the compressors discharge directly into the condensers. For this type of system, the discharge pressure (or condensation pressure) will be directly linked to the capacity and performance of the condensers and condensing fluid conditions in general.

2-stage systems. In this type of operation, booster compressors operate (compressing from low pressure to intermediate pressure) and high or second-stage compression compressors (which compress from intermediate pressure to condensation pressure, which also depends on the condensation system). The discharge pressure of the low-pressure compressors will correspond to the suction pressure of the second-stage compressors. The discharge pressure of the first-stage compressors (and the suction pressure of the second stage) is normally called intermediate pressure. This pressure will therefore depend on the operation of both the first-stage and second-stage compressors. In many refrigeration units, the intermediate pressure has been conventionally set at around 1.9 bar (corresponding to the ammonia saturation temperature of −10°C). However, this intermediate pressure would not always be the best choice for operating a refrigeration system. The intermediate pressure that will lead to the maximum COP will depend on many factors, such as evaporator capacity, equipment connected in this regime (ice machines, chilled water systems, among others), thermal load, operation of high-pressure compressors, and so on. The fact is that maintaining this fixed operating pressure may promote the best energy efficiency for few or even none of the conditions in which the system is operating. The only way to keep the system operating at a point of maximum efficiency is to implement a control with variable pressure on demand. **Figure 19** shows the consumption curve as a function of the intermediate pressure for an observed operating condition, of a refrigeration unit with 11 screw compressors and nine evaporative condensers with a nominal capacity of heat transfer rate of 16,511.6 kW, as a function of the intermediate pressure for an observed operating condition. In this case, the best operating condition resulted in an intermediate temperature around −4°C. Any other intermediate temperature will result in higher electricity consumption. It should be noted that this intermediate temperature will only be the best for this instantaneous operating condition.

Intelligent automatic control of condensing pressure is also essential for both system performance and safety. Compressors will have lower energy consumption with lower condensing pressures. Each system reacts to condensing pressure differently at different times of the day, whether due to weather conditions (wet bulb temperature – T_{WB}) or unit operating conditions (thermal load, suction pressures, individual performance of each compressor, and others).

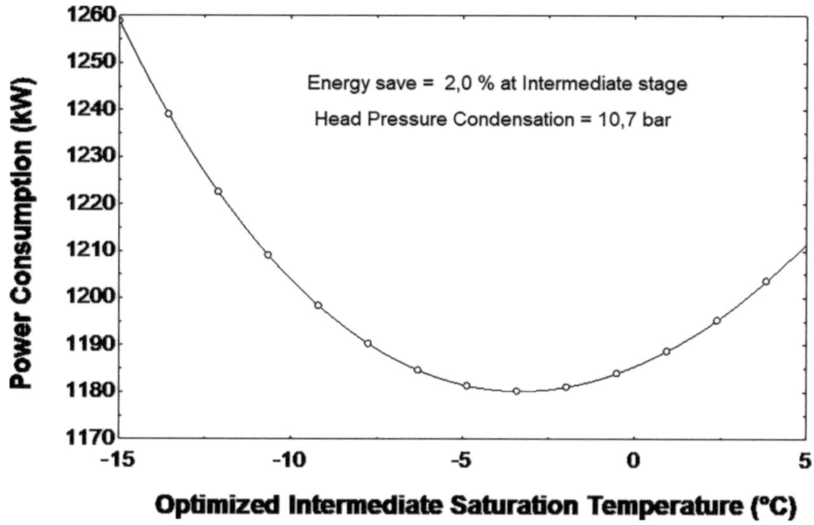

Figure 19.
Variable intermediate saturation temperature.

As an example, **Figure 20** shows the optimal condensing pressure as a function of the ambient T_{WB} for the refrigeration unit evaluated. It can be seen that for the conditions observed throughout the day, there will be several condensing pressures that will result in lower system energy consumption.

Therefore, keeping the condensing pressure constant, even if it is lower than the design values, will not provide the best performance for the system, and some energy waste will be observed.

Figure 21 shows the energy consumption of the unit evaluated for different T_{WB} values to which the system may be subjected. It can be seen from the curves in the graph that for each $T_{WB,}$ there will be an optimal condensing pressure that will lead the system to the lowest energy consumption. For both higher and lower condensing pressure values, energy consumption will be higher than it could be if it operated with an intelligent condensation control system, keeping the condensers operating at the

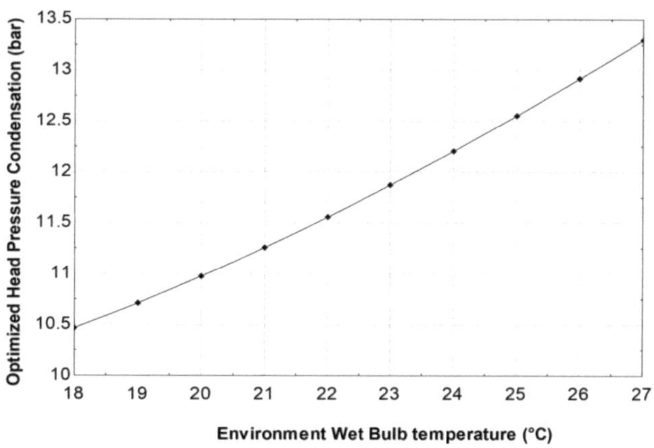

Figure 20.
Optimal condensing head pressure versus T_{WB}.

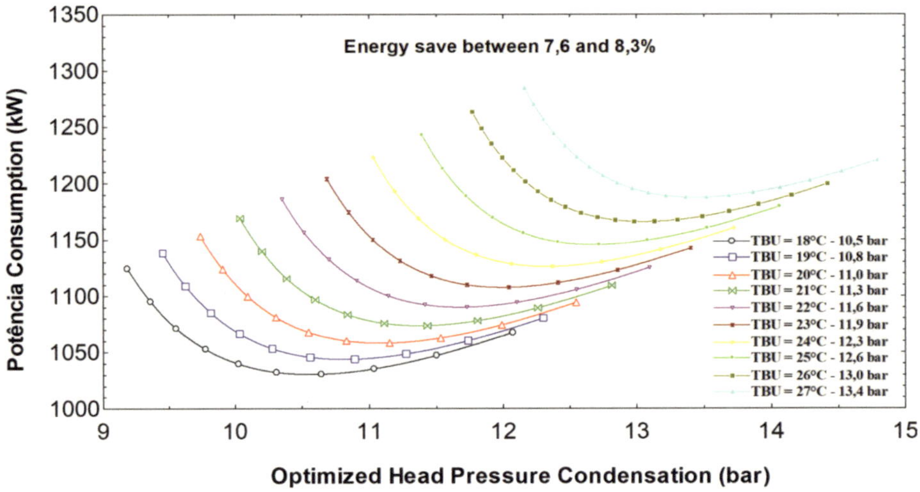

Figure 21.
Optimized condensation head pressure.

minimum consumption point. This type of system can provide significant savings in machine rooms, in addition to greater operational safety.

4. Capacity control in refrigeration screw compressors

As previously mentioned, maintaining the best operating pressures is of vital importance for the performance of a refrigeration system.

The operating pressures of compressors play a fundamental role in both the capacity and the mechanical power required to compress the refrigerant fluid. The suction pressure often corresponds to the temperature at which the fluid is pumped to the evaporators, directly influencing the capacity of this equipment and, consequently, the temperature of environments and products. The discharge pressure has a strong influence on the power required to operate the compressors and can represent something around 5 to 8% of the energy consumption for each 1 bar of pressure increase. These percentages are estimates and maybe even higher for each situation. Factors such as the efficiency of compressors and condensers will directly influence the overconsumption of this equipment.

To maintain the suction pressure of compressors, two types of strategies are normally employed: (a) mechanical drive, generally with hydraulic drive by oil pressure—slide valve; and (b) variable speed drivers (VSD). The latter is generally used in screw-type compressors.

Figure 22 shows a comparison based on manufacturer data for the variation in the capacity of a screw-type compressor by reducing capacity from 100 to 50% of its nominal capacity through variation in rotation and the compressor's slide valve.

The blue curve represents the variation in capacity with the compressor's mechanical system, while the trend that can be observed through the yellow dots refers to the reduction in capacity through variation in compressor frequency. By reducing the compressor frequency to 50%, it can be seen that the capacity reduces to less than 50%. With the mechanical device (slide valve), the capacity is greater than 50%.

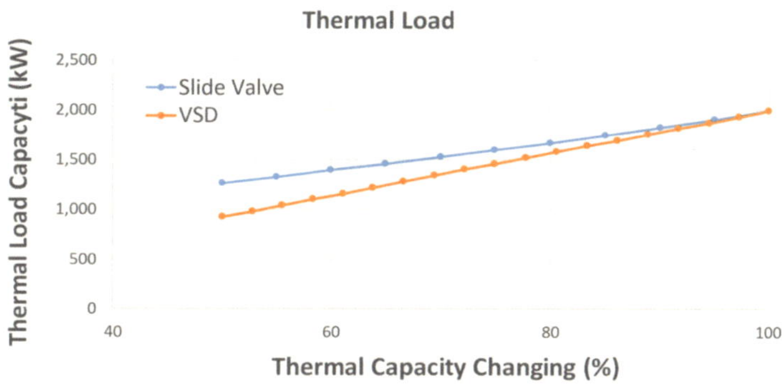

Figure 22.
Thermal capacity change in a screw compressor - VSD x Slide Valve.

The behavior of the power consumption when varying the thermal capacity of the compressor in the two ways described above is shown in **Figure 23**. Through the graph, it can be seen that by reducing the compressor rotation to 50%, the power decreases in the same proportion, while by reducing it mechanically, the powers observed will be greater.

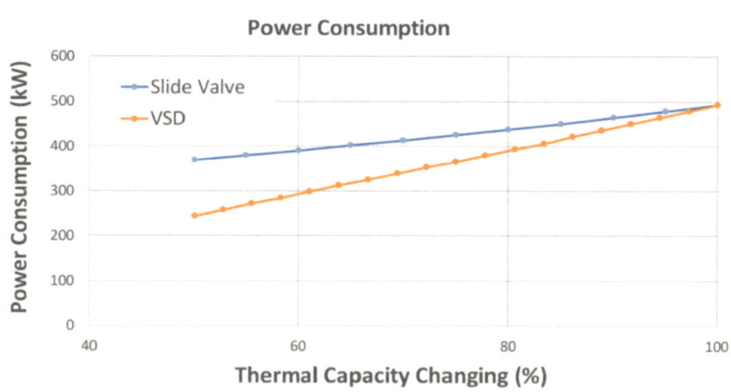

Figure 23.
Power demanded variation due to thermal capacity changing in a screw compressor - VSD x slide valve.

This analysis leads us to believe that by reducing capacity through frequency variation in a compressor, the power will be lower than through the compressor's thermal capacity control system. However, the thermal capacity will also be lower in this circumstance, and the best way to evaluate performance will be through the coefficient of performance – COP of the compressor. This parameter shown in **Figure 24** is always superior when using VSD as a control technique for compressors, according to their manufacturers. It is important to note that it will always be necessary to evaluate each situation individually since analyses performed in the field have already demonstrated trends that are opposite to those presented here. **Figure 25** shows an analysis performed on a screw compressor with a thermal capacity reduction of approximately 50%, first through the slide valve and then through the variation of the compressor speed (VSD), keeping all other compressor conditions the same. The reduction through the slide valve resulted in a current saving of approximately 136.6 A, which results in an annual energy saving of approximately 619,349.6 kWh.

In any case, the implementation of intelligent control strategies and devices can bring significant gains throughout the period in which the engine room is operating, always selecting the best setpoints for each situation.

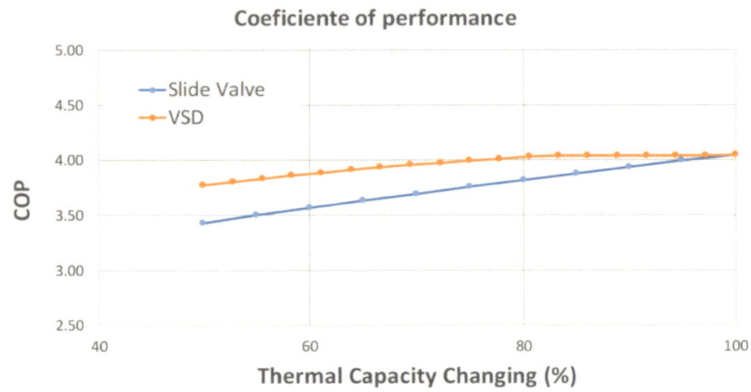

Figure 24.
COP variation due to thermal capacity changing in a screw compressor - VSD x slide valve.

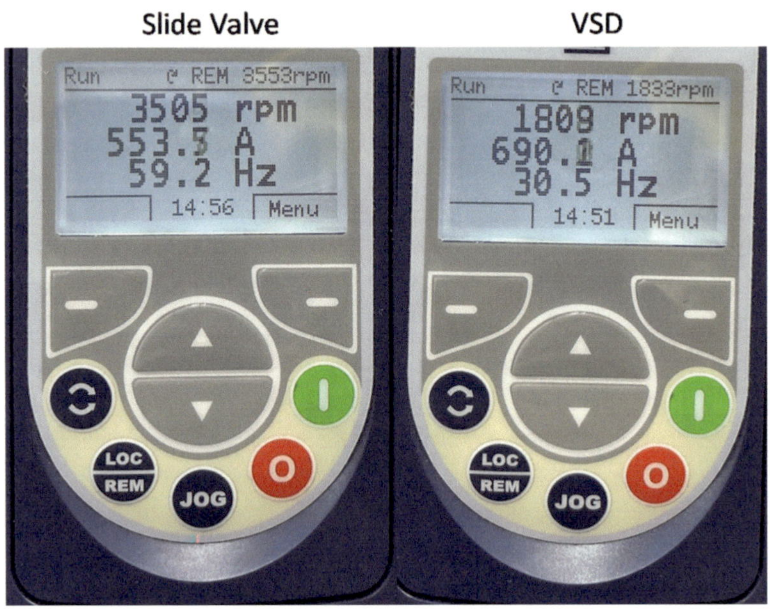

Figure 25.
Electrical current consumption by slide valve and VSD.

5. Overview about energy efficiency strategies

Therefore, this work highlights the impact of strategies and technologies employed on energy efficiency improvements at industrial refrigeration systems by bringing studies of cases about refrigeration units used in food refrigeration of dairy, pork, poultry, and beef products. The operation optimization of the ice machines, water coolers, evaporative condensers, pressures of suction and discharge of compressors, schedule of operation of compressors, and other strategies employed were used to save energy. The operational parameters are obtained by computational simulation through the mathematical models existing in the SGEM® system, which is built especially for increasing the COP to industrial refrigeration systems. **Figure 26** shows one of the interfaces of the SGEM® simulation.

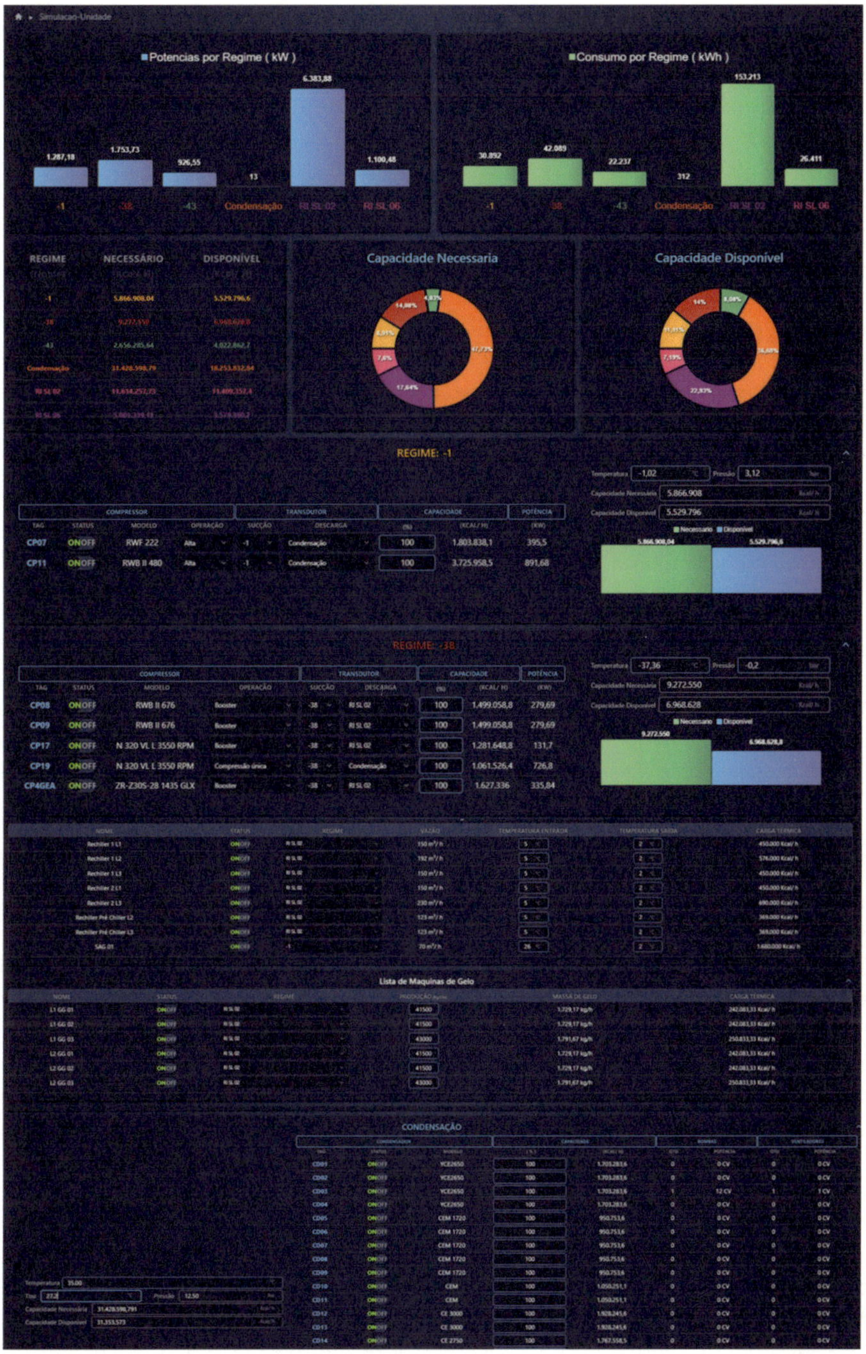

Figure 26.
SGEM computational simulator interface.

Successful energy efficiency behaviors normally start from a structure appropriate for each specificity, with the following basic procedures being recommended:

- Full knowledge of the plant;
- Updated flowcharts;
- Equipment inventory;
- Keeping the manuals up to date;
- Predictive, preventive, and corrective maintenance at the right times;
- Training of involved staff;
- Carrying out energy efficiency studies with qualified auditors;
- Mathematical modeling and computer simulation of the refrigeration unit;
- Meeting with the staff to identify critical points and opportunities;
- Definition of methodology for evaluating actions;
- Automatic activation and control of equipment;
- Equipment data measurements and comparison of operational standards;
- Remote monitoring, assessment of the current situation, and identification of opportunities;
- Continuous process assessments;

6. Conclusions

This work addressed energy efficiency in industrial refrigeration systems used for the production and storage of food and beverage products. Some strategies and technologies used to increase the coefficient of performance (COP) of refrigeration units are shown, such as intelligent operation through variable speed drives (VSDs) and intelligent controls in fans of evaporative condensers, floating head pressure work, optimization of ice and chilled water production, and use of mathematical modeling and computer simulations, among others.

Analyses of a refrigeration system for freezing poultry show an available annual energy saving of about 4,473,467.57 kWh.

Strategies indicated by this methodology can provide annual savings of approximately 648,000 kWh and an increase of 460 m^3/h of the cooled water in a refrigeration system for frozen dairy products.

The annual savings generated in a system for frozen pork products were approximately 528,000 kWh during peak hours and 1,584,000 kWh during off-peak hours, representing an improvement of approximately 5.18 kWh per ton of processed product.

Evaluation of evaporative condensers showed that restoring the best operating conditions, the annual energy saving is around 998.784 kWh in a medium-sized system, with a total condensing heat transfer rate capacity of 5528.2 kW.

One of the refrigeration units evaluated in this work showed a potential for energy savings of approximately 10% through operation with intermediate and floating condensation pressures.

Another important verification addressed in this work was the evaluation of the reduction in load thermal capacity of compressors by comparing two different ways: through a frequency inverter and a slide valve. The reduction in thermal capacity through the slide valve showed an annual energy-saving potential of approximately 619,349.6 kWh.

However, it is possible, through the results presented, to perceive the importance of strategic actions that in the vast majority of cases do not require the purchase of new equipment or trial and error methodologies that entail high costs and, in most cases, little or no significant results.

Acknowledgements

The author thanks to the financial support of the Conselho Nacioal de Desenvolvimento Citentífico e Tecnológico - Brazil (CNPq).

Nomenclature

COP	coefficient of performance
EC	evaporative condenser
PI	proportional integral
RH	relativity humidity (%)
SGEM	remote management and monitoring system
TWB	wet bulb temperature (°C)
VSD	variable speed drive

Author details

Ivoni Carlos Acunha Jr.
Institute Federal of Education, Science and Technology of Rio Grande do Sul, Rio Grande, Brazil

*Address all correspondence to: ivoniacunha@gmail.com

References

[1] Stoecker WF, Jabardo JMS. Refrigeração Industrial. 2nd ed. São Paulo: Edgard Blucher; 2002

[2] Widell KN, Eikevik T. Reducing power consumption in multi-compressor refrigeration systems. International Journal of Refrigeration. 2010;**33**(1):88-94. DOI: 10.1016/j.ijrefrig.2009.08.006

[3] Hovgaard TG, Larsen LF, Edlund K, Jørgensen JB. Model predictive control technologies for efficient and flexible power consumption in refrigeration systems. Energy. 2012;**44**:105-116

[4] Plessis GE, Liebenberg L, Mathews EH. The use of variable speed drives for cost-effective energy Savings in South African Mine Cooling Systems. Applied Energy. 2013;**111**:16-27. DOI: 10.1016/j.apenergy.2013.04.061

[5] Peng Q, Du Q. Performance evaluation of a variable frequency heat pump air conditioning system for electric bus. International Journal of Fluid Machinery and Systems. 2015;**8**(1):13-22. DOI: 10.5293/ijfms.2015.8.1.013

[6] Jin-Sik O, Binns M, Park S, Kim J-K. Improving the energy efficiency of industrial refrigeration systems. Energy. 2016;**112**:826-835. DOI: 10.1016/j.energy.2016.06.119

[7] Yin X, Wang X, Li S, Cai W. Energy-efficiency-oriented cascade control for vapor compression refrigeration cycle systems. Energy. 2016;**116**:1006-1019. DOI: 10.1016/j.energy.2016.10.059

[8] Cirera J, Carino JA, Zurita D, Ortega JA. Improving the energy efficiency of industrial refrigeration systems by means of data-driven load management. PRO. 2020;**8**(9):1106. DOI: 10.3390/pr8091106

Chapter 2

System Analysis of Thermoelectric Heat Pump Circuits with Regenerative Heat Recovery

Yuriy Lobunets

Abstract

A schematic concept of a compressorless heat pump for residential and commercial HVAC installations is considered. The technical approach is based on a solid-state thermoelectric heat pump, which will reduce energy consumption for cooling and heating compared to typical technologies. A key feature of all reverse thermodynamic cycles on which heat pumps operate is the inverse hyperbolic dependence of the coefficient of performance (COP) on the difference in operating temperatures. The considered design divides the total operating temperature difference into many small intervals, each with an increased COP. The main thing is the technical possibility of creating an integrated recuperative heat exchanger, which includes thermoelectric micro-heat pumps in a sequential heating (or cooling) circuit. Since the temperature difference on each micro heat pump is much smaller than the total, each unit works with an increased efficiency factor. In this design, the total temperature rise of the coolant is equal to the sum of the temperature gains of all individual micro heat pumps, while the efficiency is equal to their average efficiency. This arrangement of the circuit of the thermoelectric heat pump simulates the ideal Lorentz cycle with internal heat recovery, which ensures high efficiency.

Keywords: thermoelectric cooling, thermoelectric heat pump, COP, heat recovery, Lorentz cycle

1. Introduction

The EU aims to be climate neutral by 2050—an economy with zero greenhouse gas emissions. This goal is at the heart of the European Green Deal [1] and is a legally binding goal thanks to the European Climate Law [2].

Energy is central to the path to a net zero economy. The transition to a more sustainable energy system involves the abandonment of fossil fuels and the transition to low-carbon and renewable energy sources, increasing energy efficiency in products, industry and buildings, as well as creating a more sustainable energy system based on clean technologies.

Up to 50% of total energy consumption and corresponding greenhouse gas emissions are accounted for by activities related to providing comfortable thermal conditions in the utility sector. One of the promising directions for improving energy efficiency and minimizing greenhouse emissions in this sector is the transition to low-temperature heat supply systems that use renewable sources of heat energy and waste heat (the so-called 4th and 5th generation centralized heat supply systems) [3, 4]. Studies revealed that the 4th or 5th generation of district heating systems is more feasible, fuel-efficient, and cheaper than individual solutions in areas with high urban density. Such heat supply systems can become especially relevant in crisis conditions, like the current situation in Ukraine, when the stability of the energy system during the war is ensured by the decentralization of generating capacities. The fifth-generation district heating has a network with a temperature as close to the ambient ground temperature. If the DH supply temperature is 30°C or less, it cannot be used directly for space heating (SH) or domestic hot water (DHW). An electric heater or a booster heat pump is required to raise the temperature. A heat pump extracts heat from a low-temperature medium (e.g., DH supply) and delivers it to a medium at a higher temperature (e.g., building). The most common type of heat pump in modern conditions is vapor compression system heat pumps (VCC), which have an age-old history of use, but are not free from shortcomings. This applies primarily to the use of freon coolants, which are the most dangerous greenhouse gases. According to international agreements, their use is planned to be limited starting from 2025. In addition, the temperature interval of use of VCC is significantly limited by the properties of heat carriers, their performance is difficult to adjust. In this regard, the task of developing technologies for compressorless heat pumps that are free from these shortcomings is urgent. According to a study conducted by the US Department of Energy [5], among the promising technologies is the thermoelectric method of energy conversion. The advantages of thermoelectric heat pumps (THP) are high reliability, compactness, noiselessness, easy adjustment in a wide power range, and the absence of dangerous coolants. But in terms of efficiency, they are inferior to steam compression ones. This drawback is determined by the limited properties of existing thermoelectric materials. However, the features of this technology allow the use of more efficient thermal circuits and thermodynamic cycles that can bypass these limitations. This is confirmed by experimental studies that show the possibility of significantly increasing the efficiency of THP due to the use of schemes with heat regeneration [6, 7]. To determine the possible limits and conditions for improving the characteristics of THP, this work carried out a theoretical analysis of the properties of schemes with internal heat regeneration.

2. Background and research base

Thermoelectric phenomena were discovered more than 200 years ago when Seebeck discovered the effect of electromotive force in a circuit from different conductors between the contacts of which there is a temperature difference. Soon, Peltier discovered the effect of absorption and release of heat on the surface of the contacts of such a circuit during the flow of an electric current. These effects became the subject of research, during which Thomson theoretically predicted, and later experimentally proved the existence of the effect of distributed absorption of heat in conductors during the flow of current in the presence of a temperature gradient. Thanks to the works of Ioffe [8] and Goldsmith [9], it was established that the most promising thermoelectric materials are semiconductors. This marked the beginning of widespread practical use of thermoelectric effects to create generators of direct conversion of thermal energy into electrical energy and micro-refrigerators for special application conditions. Due to their high reliability, lack of moving parts, and independence from the scale factor, thermoelectric heat pumps are widely used in cooling systems and thermoregulation systems of electronic devices, in instrument and mechanical engineering. Their use in more powerful applications is restrained by relatively low efficiency.

The principle of operation of THP is based on the use of the Peltier thermoelectric effect - when a current flows through a chain of dissimilar materials, heat is absorbed in one connection, and heat is released in the other, which causes a flow of heat between the connections and a temperature difference between them. The heat balance at the joints consists of Peltier heat $Q_P = IeT$, which is compensated by the flow of thermal conductivity $Q\lambda = -\lambda gradT$ and the flow of heat exchange with the environment $Q_\alpha = \alpha \Delta T$ (**Figure 1**):

$$\begin{aligned} IeT_h &= -\lambda gradT + \alpha_h(T_h - t_h) \\ IeT_c &= -\lambda gradT + \alpha_c(t_c - T_c) \end{aligned} \tag{1}$$

Under certain circumstances, this balance can be formed in such a way that the temperature of one connection decreases below the ambient temperature $(T_c - t_c)$, and the temperature of another connection is maintained above it $(T_h > t_h)$. Thus, under the action of an electric current, heat is removed from the object of cooling at one junction of the thermocouple, and at the other, it is diverted to the surrounding environment. That is, this effect makes it possible to realize the reverse thermodynamic cycle – due to the work of an external energy source, heat can be transferred from an environment with a low temperature to a heat carrier with a higher temperature. To find the area of definition of the problem of thermoelectric cooling, a mathematical model of a thermocouple under appropriate boundary conditions is used, the basis of which is the solution of the Poisson equation of thermal conductivity with heat sources:

$$T(Y) = C_1 + C_2 Y - q_v/\lambda \, Y^2, \tag{2}$$

where integrations constant C_1, C_2 are determined depending on the boundary conditions of the problem. A. Ioffe, after considering the problem at fixed junction

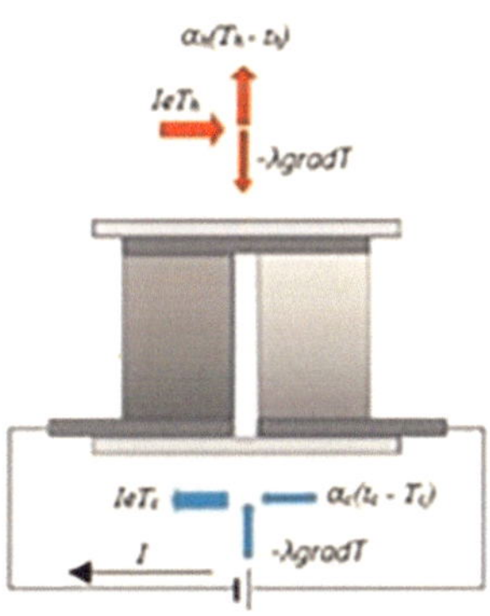

Figure 1.
Thermoelectric circuit.

temperatures [8], determined that the coefficient of performance THP (COP) can be represented as the product of the efficiency of the reverse Carnot cycle $\eta_c = T_c/\Delta T$ on the coefficient that considers losses from irreversible processes of heat conduction and Joule heat release in a thermocouple:

$$\mathrm{COP} = \eta_c\,(M - T_h/T_c)/(M+1), \tag{3}$$

where $M = \sqrt{(1 + zT)}$ is a parameter that depends only on the dimensionless criterion of the thermoelectric figure of merit of the material zT;

$z = e^2\sigma/\lambda$ – thermoelectric figure of merit, K^{-1};

$T = (T_h + T_c)/2$ is the average temperature of the junctions of the thermocouple.

This prompted researchers to search for materials with high zT. Even though after a period of rapid growth, the progress in improving the properties of materials slowed down and reached the level of saturation, this direction is still considered the main one for solving the problem of THP efficiency. At the same time, there is an opinion that there are thermodynamic limitations on the growth of zT, and the maximum possible value of the thermoelectric factor is limited by the condition $lim(zT) \to 1$, [10]. The latter assumption conflicts with a significant number of reports on the development of materials with $zT > 1$. However, these results for the most part do not ensure reproducibility and distribution from laboratory samples of materials to specific products [11–13]. At the same time, the limited thermoelectric figure of merit of materials does not mean that thermoelectrics are unable to provide a competitive level of efficiency for energy applications. Most of the known estimates of the efficiency of THP are based on the analysis of the relative internal efficiency of the thermal energy converter, that is, only internal irreversible losses in thermocouples are considered. However, the energy conversion factor COP = Q/N is not an absolute criterion for the efficiency of the energy source. In fact, a more objective indicator of the economic efficiency of a heat pump is the levelized cost of thermal energy (LCOH). This indicator is a complex function of various factors and does not depend uniquely only on the thermoelectric figure of merit. The levelized cost of heating estimates the average cost of providing 1 KJ of heating over the lifetime of the equipment, considering the capital cost of the equipment and installation; operating expenses include the cost of fuel and regular maintenance [14]. Economic efficiency depends on thermal efficiency, which considers external irreversible losses of temperature pressure, the conditions of heat exchange, the cost of THP components, the price of energy carriers, and so on. According to Eq. (3), COP significantly depends on the temperature difference between the junctions ΔT. A general property of reverse thermodynamic cycles is the hyperbolic dependence of the efficiency on the difference in operating temperatures: at $\Delta T \to 0$, the COP increases indefinitely (**Figure 2**).

Also, the characteristics of THP are significantly affected by the heat exchange at the junctions of thermocouples. Irreversible losses from external heat exchange increase the temperature difference between the junctions, which is necessary to achieve the required cooling depth, which leads to a drop in COP compared to the ideal model. Many works [15–17] are devoted to the analysis of THP regimes in various conditions. In a generalized form, the solution to the problem of temperature distribution in thermocouples can be presented using a system of dimensionless criteria that have a clear physical meaning and standardized change intervals, [18]:

$$\Theta = f\left(\mathrm{zT}_p, J, \mathrm{Ki}, \mathrm{Bi}\right), \tag{4}$$

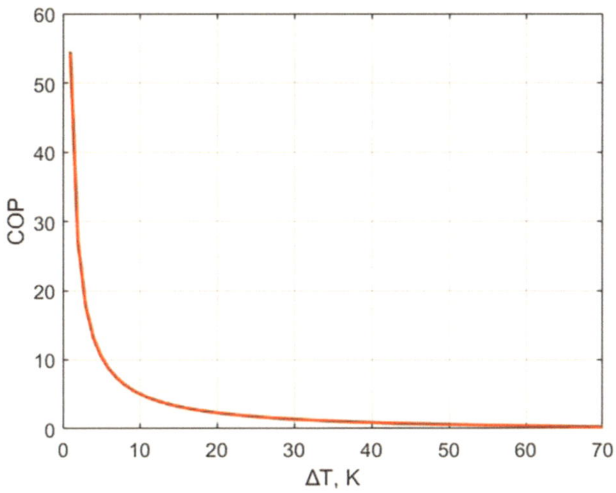

Figure 2.
Dependence of the efficiency of THP on the temperature difference.

Here $\Theta = T/T_p$ is the dimensionless temperature of the thermoelement, which is used to calculate the energy characteristics of the THP; zT_p is a criterion of thermoelectric figure of merit that characterizes the thermoelectric material. Dimensionless current $J = jeh/\lambda$, dimensionless heat flow $\mathrm{Ki} = qh/\lambda T_p$, and Bio's criterion $\mathrm{Bi} = \alpha h/\lambda$ characterize the mode parameters that determine the state of the system. The defining temperature of the problem T_p, which sets the temperature scale, is equal to the ambient temperature $T_p = t_o$.

To consider external irreversible losses during heat exchange, a mathematical model is used in boundary conditions of the third kind (Newton-Robin problem):

$$\Theta\,(Y) = C_1 + C_2Y\text{–}J^2/\mathrm{zT}\;Y^2$$

$$\mathrm{Bi}_c[\vartheta_c - \Theta(0\,)] + \Theta'(0)\text{–}\mathrm{J}\Theta(0) = 0; \tag{5}$$

$$\mathrm{Bi}_h[\Theta(1\,) - \vartheta_h] + \Theta'(1)\text{–}\mathrm{J}\Theta(1) = 0;$$

Constant integrations C_1, C_2 in this case are determined from the system of (Eq. (5)) as:

$$C_1 = C_F/C_L;$$
$$C_2 = C_1{}^*\;(J + \mathrm{Bi}_c) - \mathrm{Bi}_c{}^*\,\vartheta_c, \tag{6}$$

where

$$C_F = \mathrm{Bi}_h{}^*\;\vartheta_h + J^2/\mathrm{zT}^*\;(1 + (\mathrm{Bi}_h - J)/2) + \mathrm{Bi}_c{}^*\,\vartheta_c{}^*\,(\mathrm{Bi}_h\text{–}J + 1);$$
$$C_L = \mathrm{Bi}_h\text{–}J + (J + \mathrm{Bi}_c)^*\,(\mathrm{Bi}_h\text{–}J + 1);$$

In the general case, the heat flux at the heat-absorbing joint is equal to:

$$q = \alpha_c(t_c - T_c), \tag{7}$$

or in dimensionless form:

$$\mathrm{Ki} = \mathrm{Bi}_c\ (\vartheta_c - \Theta_c) \tag{8}$$

It follows from the last relation that the heat pump mode is realized if the temperature of the cold junction is lower than the temperature of the object being cooled.

$$\Theta_c \le \vartheta_c \tag{9}$$

For arbitrary combinations of problem parameters, the range of permissible supply currents $J_{min} \le J \le J_{max}$ satisfying condition (Eq. (9)) is found as the roots of the equation

$$\vartheta_c - \Theta_c\left(J, \mathrm{zT}_p, \mathrm{Bi}_c, \mathrm{Bi}_h, \vartheta_c, \vartheta_h\right) = 0; \tag{10}$$

At the ends of this range of currents, the temperature of the heat-absorbing joint is equal to the temperature of the object being cooled, and the cooling capacity turns to zero, **Figure 3**. The relationship between the cooling capacity and the efficiency of the standard thermoelectric module for the given temperatures of the heat source and sink is illustrated in **Figure 4**. A comparison of the results of the real and idealized model indicates a significant influence of the heat exchange conditions on the characteristics of the THP. The analysis of expression (Eq. (10)) proves that for any specific conditions, there is a threshold of the intensity of heat exchange, below which the interval of admissible power currents turns to zero, and the implementation of the cooling mode is impossible.

The given data prove that the efficiency and power of the THP in each temperature range can be adjusted within wide limits, which makes it possible to determine the optimal modes that provide the best technical and economic characteristics of the

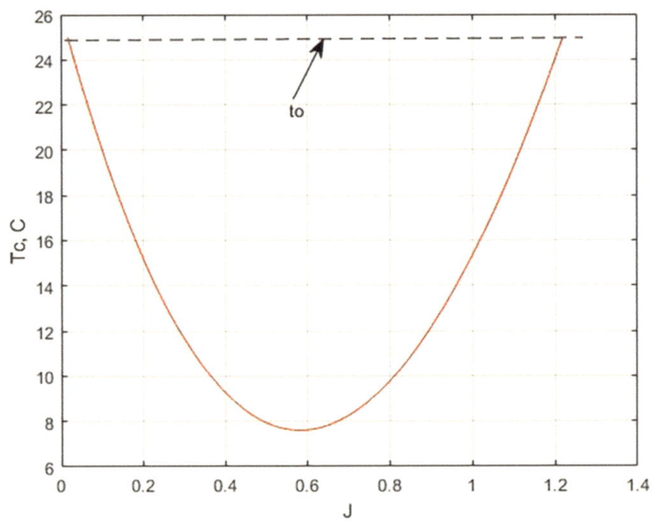

Figure 3.
Dependence of the temperature of the cold contact $\mathrm{T_c}$ *on the supply current* J.

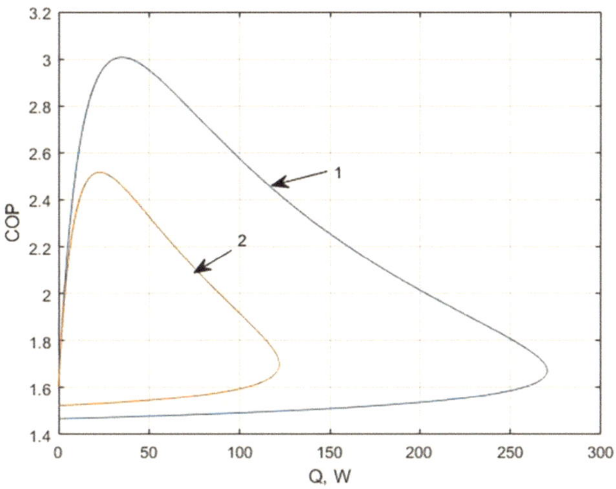

Figure 4.
Correlation between cooling capacity and efficiency of the thermoelectric module. 1: an idealized model; 2: an real model.

THP, and to develop control algorithms capable of adapting the THP mode to current conditions.

The considered results concern low-power THPs with constant temperatures along cold and hot junctions. That is, in such devices, the reverse thermodynamic Carnot cycle between two isotherms is implemented. In more powerful THPs, heat transfer is carried out by coolants that are cooled/heated, flowing along the surface of the thermoelectric modules. At the same time, the temperatures of the coolants, which play the role of the heat source and heat sink, change (**Figure 5**). In fact, such a heat pump is a recuperative heat exchanger that provides heating and cooling of coolants due to thermoelectric transfer of heat between them. The presence of intermediate coolants makes it possible to increase the efficiency of THP due to the organization of schemes with heat regeneration and recirculation of coolants. Since the heating and cooling are carried out at alternating temperatures of the heat source and sink, this process is closer to the reverse Lorentz thermodynamic cycle. It is known that the efficiency of this cycle is at least twice as high as the efficiency of the Carnot cycle in the same temperature ranges [19]. This is because, in the case of non-isothermal heat source and heat sink, the Lorentz thermodynamic cycle can be represented as a set of separate isothermal Carnot cycles, the operating temperature difference of each of which $(T_h - T_c)$ can be significantly smaller than the total operating temperature interval $(t_{ho} - t_{co})$. That is, each local cycle works in more favourable conditions, which causes an increase in the overall efficiency of the cycle. In such a scheme, many serially connected thermoelectric elements allows for dividing the total difference in operating temperatures into many small intervals. Thus, each micro-THP has a high COP. On this basis, it is technically possible to create an integrated recuperative heat exchanger that combines thermoelectric micro-heat pumps into a sequential cooling/heating circuit, along which the temperature of coolants circulating in a countercurrent circuit in the circuit of the heat source and the circuit of the consumer changes sequentially. Since the temperature difference on an individual micro heat pump can be very small (≈5°C), each of them works with a high-efficiency factor. The operating temperature difference of the THP is equal to the sum of the temperature increases of the coolant on all individual micro heat pumps, while the overall efficiency is

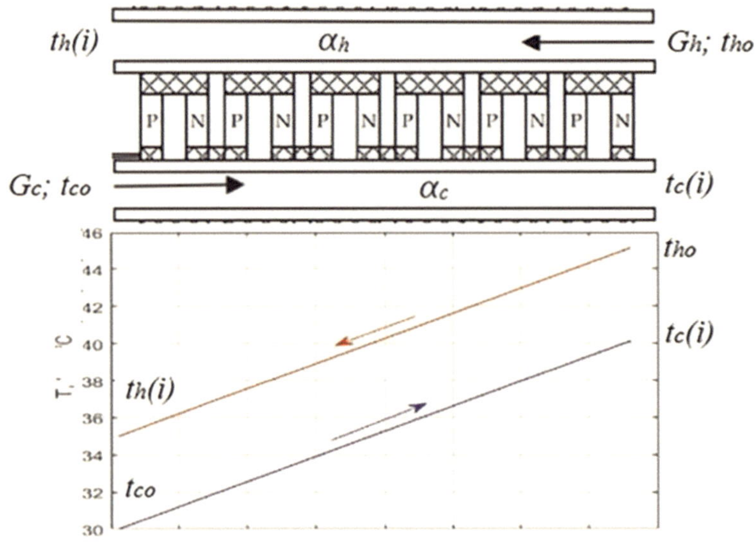

Figure 5.
Scheme of THP heat exchange type.

determined by the efficiency of individual units. Such a thermoelectric heat pump can work in any range of operating temperatures since the output temperatures can be adjusted due to the number of serially connected thermocouples and the supply current. This, together with such factors as the absence of refrigerants, flexible power regulation, and compatibility with digital control systems, opens the prospects for the wide use of the considered scheme of thermoelectric heat pumps in local HVAC systems.

3. Subject and methodology of analysis

The configuration of the studied system, which includes a thermal point of the central heating system with a booster heat pump, is shown in **Figure 6**.

This is a so-called direct DH substation, which does not have a heat exchanger that separates the primary flow from the secondary flow. The heat carrier from the central DH system enters directly into the local network for space heating and domestic hot water preparation (DHW). Water with t_{ho} temperature enters the consumer's heating system from the low-temperature central heating system. For use in the space heating and hot water supply system, its temperature must be increased to t_{h1}. For this, the incoming water of the central heating system is fed to the booster heat pump, where it is heated to the required temperature, and enters the internal heat supply system. Here it gives heat Q_c to consumers and cools down to the temperature t_{h2}. After that, it returns to the heat pump, where it transfers heat Q_R to the direct flow of the heat carrier and enters the central heating system with a temperature of t_{c2}. That is, in such a scheme, internal heat regeneration is carried out and the initial temperature of the heat source for THP rises from t_{ho} to t_{h1}. The supply current J_o is chosen in such a way as to ensure an increase in the temperature of the coolant from t_{ho} to t_{h1} in the THP supply channel, which consists of n thermoelectric modules.

The heat balance in this system is as follows:

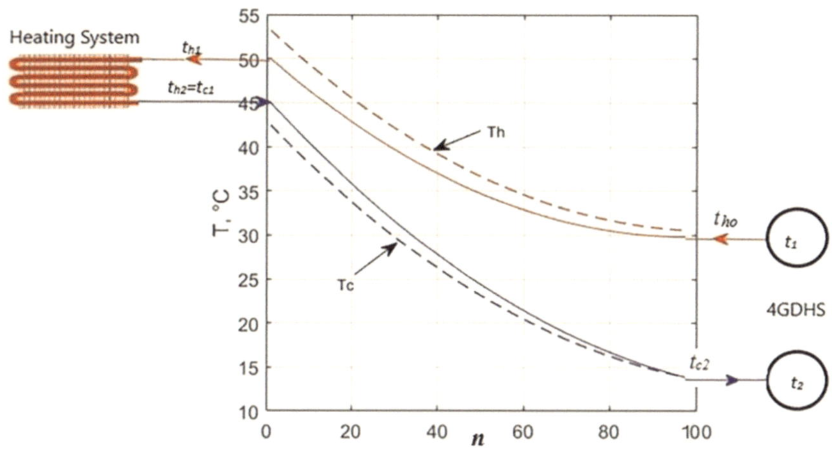

Figure 6.
Scheme of heat supply from THP.

$$Q_o = Q_c + Q_R, \tag{11}$$

where $Q_o = W(t_{h1} - t_{c2})$ is the total amount of consumed heat; $Q_c = W(t_{h1} - t_{h2})$ is the heat consumed by the heating system; $Q_R = W(t_{c1} - t_{c2})$ is the heat of internal regeneration.

THP consumes power equal to the sum of the powers of individual modules:

$$N = \text{sum}(N(i)), \tag{12}$$

where $N(i) = [J_o^2/zT_p + J_o\Delta\Theta(i)]\lambda T_p/h$ is the power of a separate module, W; $\Delta\Theta(i)$ is the dimensionless temperature difference on the module.

Accordingly, the efficiency of THP is equal to

$$\text{COP} = Q_o/N \tag{13}$$

The purpose of the analysis is to determine the optimal parameters of similar THP schemes and to assess possible technical and economic characteristics in the conditions of low-temperature heating systems. To calculate the characteristics of the THP, it is necessary to determine the optimal supply current J_o, as well as the temperature distribution of the coolant along the channels and the temperature distribution in the thermocouples. The calculation algorithm begins with determining the permissible current in the initial temperature interval $dt = t_{h1} - t_{ho}$ according to Eq. (10). After that, the procedure for calculating the temperature distribution in the system is performed in the following sequence:

- with the help of solution (Eqs. (4) and (5)) for the given temperatures of heat carriers t_{h1}, t_{co}, the temperatures T_h, T_c of the first module, as well as the heat flows on the surfaces of the module Q_h, Q_c are calculated;

- this allows you to determine the temperature increase of the coolant and the boundary conditions for the following modules as

$$t_c(i+1) = t_c(i) - Q_c i/W_c; t_h(i+1) = t_h(i) - Q_h i/W_h.$$

The supply current, which satisfies the conditions of the problem, is selected with the help of this algorithm from the permissible interval under the condition that the calculated temperature of the coolant at the entrance the heat exchanger $t_h(n)$ is equal to the supply temperature t_{ho}, that is, as the root of the equation

$$t_{\text{ho}} - t_h(n) = 0 \qquad (14)$$

Then, according to Eqs. (11)–(13), thermal and electric power and COP THP are determined.

This algorithm makes it possible to investigate the influence of various factors on the efficiency of THP and to estimate the limits of changes in THP efficiency in the space of the main determining parameters. The implementation of the algorithm and the analysis of the problem were carried out in the environment of the MATLAB software package.

4. Results and discussion

The basic set of initial parameters typical for low-temperature centralized heating systems [6], is used for the analysis:

coolant—water;

coolant temperature at the entrance to the THP t_{ho} = 30°C;

coolant temperature at the exit from the THP t_{h1} = 42°C;

coolant temperature at the exit from the heating system t_{h2} = 37°C;

mass consumption of the coolant G = 0.001–0.003 kg/sec;

coefficient of heat transfer in channels α = 0.5–2 W/cm^2 K;

number of thermoelectric modules along the channel n = 10–100;

The choice of the mass consumption of the heat carrier corresponds to the power of the THP in the range of 1–5 kW. The chosen intensity of heat exchange is typical for plate or mini-channel heat exchangers in the considered conditions.

Properties of thermoelectric modules.

Type MT2,6–0,8–263.

The size is 50 × 50 mm.

The number of thermoelements is 526.

The dimensions of the thermoelements are 0.161 × 0.161 × 0.8 mm.

Seebeck coefficient, e (V/K).

Thermal conductivity coefficient λ, W/cm^2 K.

Coefficient of electrical conductivity σ, $[\Omega\text{cm}]^{-1}$.

Thermoelectric properties (e, σ, λ) are described with sufficient accuracy as a function of temperature by polynomials of the 2nd degree.

The determining factor affecting the efficiency of the THP is the temperature difference on the thermocouples. According to the mathematical model (Eqs. (5) and (6)), the distribution of temperatures in THP depends on the conditions of heat exchange (criterion Bi), temperatures of coolants (t_{ho}, t_{h1}, t_{co}), thermoelectric properties of the material (zT_p), supply current (J) and geometric dimensions of the device. The power of THP largely depends on the conditions of heat exchange between heat carriers. According to the theory of heat transfer [20], in countercurrent heat exchangers in the case of equal heat capacity of the mass flow of the coolants $W_h = W_c$, the temperature distribution along the heat exchanger is linear, and the temperature difference between the heat carriers is constant. It depends only on the ratio of the thermal resistance of the heat exchanger $Rt = 1/kF$ and the heat capacity of the coolant W:

$$t_h(i)\text{–}t_c(i) = \Delta t_o/(1 + \mathrm{kF}/W) \tag{15}$$

where k is the heat transfer coefficient, W/cm^2 K; F is the heat exchange surface, cm^2.

k is determined by the conditions of heat exchange in the channels and the thermal resistance of the thermoelements $R_t = h/\lambda$; F depends on the number of thermoelectric modules in the channel; $W = GC_p$ is determined by the mass flow rate of the coolant.

In a thermoelectric heat exchanger, due to the action of internal and surface heat sources (Joule and Peltier effects), relation (Eq. (15)) is violated, but the influence of these parameters remains predominant. That is, the independent parameters of the problem are the properties and number of thermoelectric modules, and the mass flow rate of the coolant, which are combined in the form of dimensionless complexes kF/W, and the supply current J. The dependence of the temperature drop on the thermoelements on these parameters is shown in **Figure 7**. The dependence of the THP efficiency on the mode parameters (**Figure 8**) illustrates the possibility of achieving COP values approaching the level of vapor compression heat pumps [21]. The ratio of efficiency and power of the considered THP scheme is shown in **Figure 9**. The presented data confirm the possibility of flexible control of THP parameters in a wide range of capacities by adjusting mode parameters. The determining parameter KF/W with a fixed THP design (channel length n = const) can vary mainly due to the mass flow rate of the coolant G. At the same time, the maximum allowable mass flow rate, beyond which condition (Eq. (14)) cannot be fulfilled, depends on the intensity of heat exchange in the channels and interval of permissible currents J. Since the intensity of heat exchange, in turn, depends on the flow rate of the coolant, these

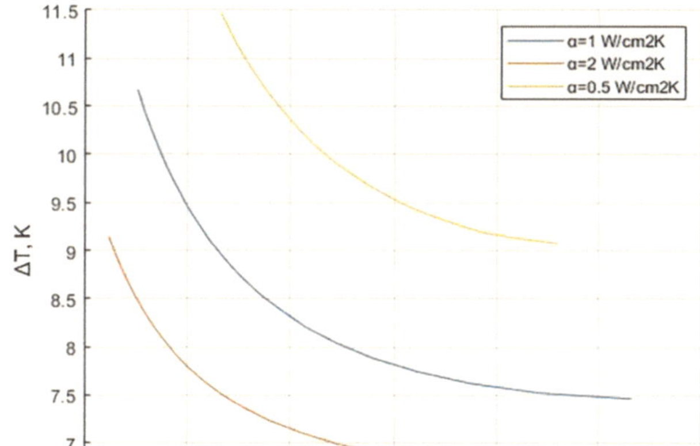

Figure 7.
Dependence of the temperature difference on the thermoelements on the ratio of the heat transfer capacity of THP KF and the mass flow rate of the heat carrier.

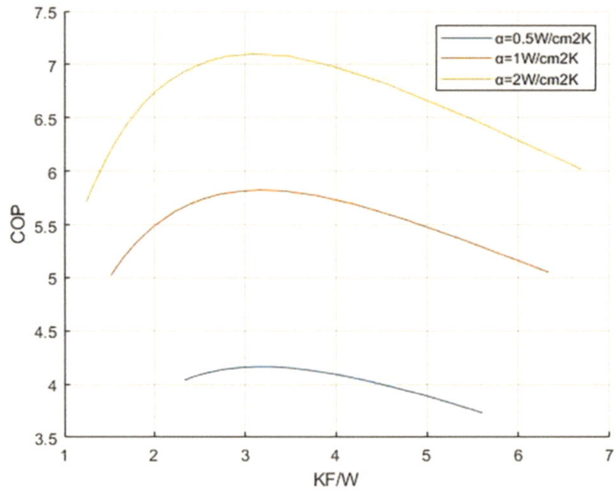

Figure 8.
Dependence of the efficiency COP on the ratio of the heat transfer capacity of THP KF and the mass flow rate of the heat carrier W.

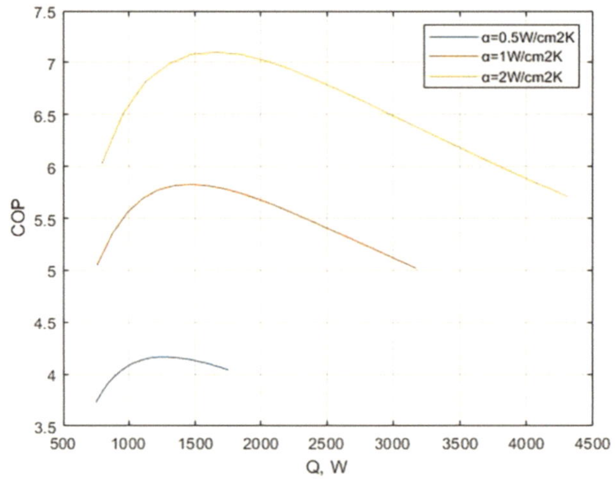

Figure 9.
Dependence of useful power Q and THP efficiency COP on heat exchange intensity α.

parameters are interrelated. That is, for a specific THP design, algorithms can be developed that automatically optimize the THP parameters in relation to the required load by controlling the flow of the coolant and the supply current.

Within the scope of the task definition, the THP with the optimal number of modules (50 units) provides power within the Q_o = (1–4) kW with a maximum COP≈ 7.

A detailed levelized cost of heat (LCOH) was examined using the techno-economic data available in the reviewed studies. According to [22]:

$$LCOH = \frac{Io + \sum_{1}^{t} \frac{At}{1+r}}{\sum_{1}^{t} \frac{Qs}{1+r}} \qquad (16)$$

TE module cost including mounting and connectors parts	\$20
Number of TE modules, *n*	50
Heat exchangers cost	\$150
Assembling cost	\$500
System THP cost, I_o	\$1650
levelized cost of electricity, LCOE	0.15 \$/kWh, [23]
The cost of electricity	Pc = LCOE*Ni, \$/Year
Maintaining cost	P_m = 100, \$/Year
O&M expenditures	$A_t = P_c + t_g P_m$, \$/Year
discount rate, r	3%

Table 1.
Input data for calculating the levelized cost of heat.

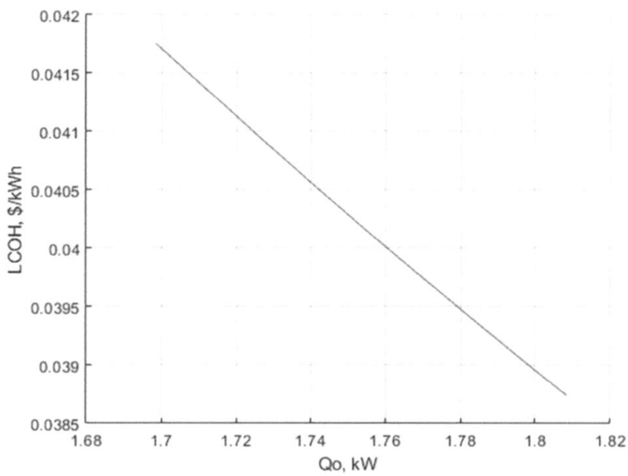

Figure 10.
Dependence of LCOH on the nominal capacity of THP.

where:

LCOH- levelized cost of heat [\$/kWh];

I_o—specific THP system costs incl. Installation [\$];

A_t—fixed and variable O&M expenditures in the year *t*, [\$];

Q—heat energy yield in the year *t* [kWh].

r—discount rate in %.

t—year within the period of use [1, 2, ... t_g].

t_g—period of use (THP thermal system lifetime in years), [10 years];

Q_s—heat produced in year *t*: $Q_s = Q_o \times 365$ day $\times$ 24 h, [kWh];

The data given in **Table 1** were used for the analysis:

The cost of supplying the coolant from the central heating network is not included in the operating costs, since this parameter depends on many external factors that are not characteristic of the analyzed scheme. That is, the obtained cost data relate only to the assessment of the operating conditions of the booster heat pump. The dependence of *LCOH* on the nominal capacity of THP is shown in **Figure 10**. Considering the fact that for traditional DH systems, *LCOH* is from 0.05 to 0.30 \$/kWh [24], the obtained results can be considered quite acceptable for the local heat supply systems.

5. Conclusion

In this chapter, the scheme of the booster thermoelectric heat pump for low-temperature central heating systems is considered. Booster heat pumps are used in LTDH to ensure the required temperature level of the heat carrier at the entrance to the local network of the final consumer. The advantage of the THP cycle in question is the use of a direct heat carrier supply scheme, without an intermediate heat exchanger. This makes it possible to ensure internal regeneration of heat in the cycle and increase the energy efficiency to the level of COP ≈ 7, which exceeds the maximum efficiency of vapor compression heat pumps for the given conditions (COP ≈ 5, [21]). A theoretical analysis of the influence of determining parameters on the characteristics of the THP scheme under consideration was carried out. The levelized cost of thermal energy under LTDH conditions was estimated. It is shown that in optimal operation modes, THPs can provide acceptable economic efficiency (*LCOH*≈ 0.04 \$/kWh), which is not inferior to traditional schemes (0.05 < *LCOH* < 0.5 \$/kWh, [24]). Further theoretical and experimental research is needed to specify the obtained results in order to develop a practical design for the THP scheme in question, verify the results, and outline prospective areas of application.

Acknowledgements

Research was carried out in the institutions of the Academy of Sciences of Ukraine in 2022–2024 according to the target program of scientific research No. 1230.

Conflict of interest

The authors declare no conflict of interest.

Acronyms and abbreviations

DH	district-heating system
LTDH	low-temperature district heating system
HVAC	heating, ventilation and air conditions system
SH	space heating
DHW	domestic hot water
HP	heat pump
VCC	vapor compression heat pump
THP	thermoelectric heat pump
LCOE	levelized cost of electricity
LCOH	levelized cost of heat

Nomenclature

Bi	Biot criterion
I	electrical current (A)
j	current density (A/cm^2)
e	Seebeck coefficient (V/K)
E	electromotive force, EMF (V)
λ	coefficient of thermal conductivity (W/cm_K)
σ	coefficient of electrical conductivity, $[\Omega \text{cm}]^{-1}$
h	thermocouple leg length (cm, mm)
$J = jeh/\lambda$	dimensionless current density
n	the charge carrier concentration (cm^{-3})
s	thermoelectric leg cross sectional area (cm^2, mm^2)
T_p	determining temperature (K)
T_h	hot junction temperature (K)
T_c	cold junction temperature (K)
ΔT	junction temperature difference (K)
t_h	heat carrier temperature (K)
t_c	coolant temperature (K)
dt	temperature difference of heat carriers (K)
$\theta = T/T_p$	dimensionless temperature
$\vartheta = t/T_p$	dimensionless temperature of fluid
$z = e^2\sigma/\lambda$	thermoelectric figure-of-merit (K^{-1})
zT_p	dimensionless thermoelectric figure-of-merit
N	electrical power (W)
N_o	specific power (W/cm^2)
$N_x = Nh/(\lambda s T_p)$	dimensionless power
q	heat flow density (W/cm^2)
Q	heat flow (W)
$Ki = qh/\lambda T_p$	dimensionless heat flow density
η	energy conversion factor/efficiency
$\eta_c = (T_h - T_c)/T_h$	Carnot efficiency
COP	THP efficiency (coefficient of performance)
α	heat transfer coefficient (W/cm^2_K)
R_t	thermal resistance (K cm^2/W)
R	electrical resistance (Ω)
$Y = y/h$	dimensionless coordinate
$W = GC_p$	heat capacity of the mass flow of the coolant (W/K)
G	mass flow of the coolant (kg/sec)
C_p	specific heat capacity (J/kg K)
Indices	*c*—cold; *h*—hot

Author details

Yuriy Lobunets
Institute of Renewable Energy, Kyiv, Ukraine

*Address all correspondence to: yurilobunets@yahoo.co.uk

References

[1] The European Green Deal. Available from: https://commission.europa.eu/strategy-and-policy/priorities-2019-2024/european-green-deal_en

[2] The European Climate Law. Available from: https://climate.ec.europa.eu/eu-action/european-climate-law_en

[3] Lund H, Østergaard PA, Nielsen TB, Werner S, Thorsen JE, Gudmundsson O, et al. Perspectives on fourth and fifth generation district heating. Energy. 2021;**227**:1-7. DOI: 10.1016/j.energy.2021.120520

[4] Khosravy M. Recent Progress in District Heating with Emphasis on Low-Temperature Systems. London, UK: IntechOpen; 2020. Available from: https://www.intechopen.com/chapters/74667

[5] Goetzler W, Zogg R, Young J, Johnson C. Energy Savings Potential and RD & D Opportunities for Non-Vapor-Compression HVAC Technologies. U.S. Department of Energy; 2014. DOI: 10.2172/1220817

[6] Tomc U, Nosan S, Vidrih B, Bogić S, Navickaite K, Vozel K, et al. Small demonstrator of a thermoelectric heat-pump booster for an ultra-low-temperature district-heating substation. Applied Energy. 2024;**361**(1):122899. DOI: 10.1016/j.apenergy.2024.122899

[7] Hou B, Zheng Y, Xing L, Song Q. Performance of a thermoelectric heat pump with recirculation and regenerative heat recovery. Applied Thermal Engineering. 2023;**120042**. DOI: 10.1016/j.applthermaleng.2023.120042

[8] Ioffe AF. L Semiconductor Thermoelements and Thermoelectric Cooling. London: Infosearch; 1957. p. 184

[9] Julian Goldsmith H. Introduction to Thermoelectricity. Berlin, Heidelberg: Springer; 2009. p. 242

[10] Lobunets Y. ZT Theorem of Thermoelectricity. SSRN Electronic Journal. 2023:1-8. DOI: 10.2139/ssrn.4363141

[11] Jeffrey Snyder G, Toberer ES. Complex thermoelectric materials. Nature Materials. 2008;7:10-14

[12] U.S. Energy Information Administration. Levelized Costs of New Generation Resources in the Annual Energy Outlook 2022. 2022

[13] Yee SK, LeBlanc S, Goodson KE, Dames C. $ per W metrics for thermoelectric power generation: Beyond ZT. Energy & Environmental Science. 2013;**6**:2561-2571. DOI: 10.1039/c3ee41504j

[14] Yoann L et al. Description: Guideline for Levelized Cost of heat (LCoH) Calculations for Solar Thermal Applications. International Energy Agency; 2017

[15] Zhao D, Tan G. A review of thermoelectric cooling: Materials, modeling and applications. Applied Thermal Engineering. 2014;**66**. DOI: 10.1016/j.applthermaleng.2014.01.074

[16] Apertet Y, Ouerdane H, Michot A, Goupil C, Lecoeur P. On the efficiency at maximum cooling power. EPL (Europhysics Letters). 2013;**103**(4):1-6. DOI: 10.1209/0295-5075/103/40001

[17] Kaushik SC, Manikandan S, Hans R. Energy and exergy analysis of thermoelectric heat pump system.

International Journal of Heat and Mass Transfer. 2015;**86**:843-852. DOI: 10.1016/j.ijheatmasstransfer.2015.03.069

[18] Lobunets Y. On the criteria for evaluating the characteristics of thermoelectric energy converters. Thermoelectricity. 2014;**2**:68-82

[19] Martynovsky VS. Cycles, Schemes and Characteristics of Thermal Transformers. Moscow: Energiya; 1979. p. 288

[20] Hausen H. Warmeubertragung im Gegenstrom, Gleichstrom und Kreuzstrom. Berlin-Heidelberg-New York: Springer Verlag; 1976

[21] Domanski PA, Brown JS, Heo J, Wojtusiak J, McLinden MO. A thermodynamic analysis of refrigerants: Performance limits of the vapour compression cycle. International Journal of Refrigeration. 2014;**38**:71-79

[22] Veynandt F, Ramschak T, Louvet Y, Köhl M, Fischer S. LCoH Calculation Method: Comparison between Task 54 and Solar Heat World Wide. TA$K54, INFO Sheet A01. International Energy Agency; 2017

[23] Lazard's Levelized Cost of Energy Analysis—Version 16.0. Available from: https://www.lazard.com/research-insights/2023-levelized-cost-of-energyplus/

[24] IEA. Levelized Cost of Heating (LCOH) for Consumers, for Selected Space and Water Heating Technologies and Countries. Paris: IEA; 2021. Available from: https://www.iea.org/data-and-statistics/charts/levelized-cost-of-heating-lcoh-for-consumers-for-selected-space-and-water-heating-technologies-and-countries

Chapter 3

Adiabatic and No-Adiabatic Absorption Chillers Using Ammonia-Lithium Nitrate Solutions

Alejandro Zacarías, Guerlin Romage, Cuauhtémoc Jiménez, José de Jesús Rubio, Ignacio Carvajal and María Venegas

Abstract

One of the applications of ammonia has been the use of this element as a refrigerant in air conditioning systems for buildings, process industries and residential homes. The high electrical consumption of conventional chillers is one of the main reasons why alternative cooling systems have been sought. This chapter presents the analysis and modeling using energy and mass balances, as well as the simulation of absorption chillers. In addition, the analysis and discussion of the results of performance parameters of adiabatic and non-adiabatic absorption chillers using the lithium-ammonia-nitrate solution are presented. From the results presented in this section, it can be seen that although the heat flow and circulation ratio are higher and the cop is lower for absorption chillers, when the condensation temperature is lower and the evaporation and generation temperatures are higher, these design parameters are closer to those of non-adiabatic absorption chillers.

Keywords: absorption refrigeration chillers, adiabatic absorption, ammonia-lithium nitrate solutions, sustainable refrigeration, solar cooling

1. Introduction

One of the applications of ammonia has been the use of this element as a refrigerant in air conditioning systems for buildings, process industries and residential homes. An air conditioning system, as seen in **Figure 1**, is made up of a system of ducts and spread; the air handling unit that contains the heating and cooling coils, and the chiller. The latter operates with a working fluid called refrigerant and can be organic or natural, such as ammonia. In the figure, outside air is forced by the fan of the air handling unit through the cooling coil and duct system to supply it into the conditioned space. The cooling coil is fed by the chiller.

1.1 Ammonia compression chillers

The chillers operate in a vapor compression refrigeration cycle, such as the one shown in **Figure 2**. The heat from the space to be refrigerated is absorbed in the

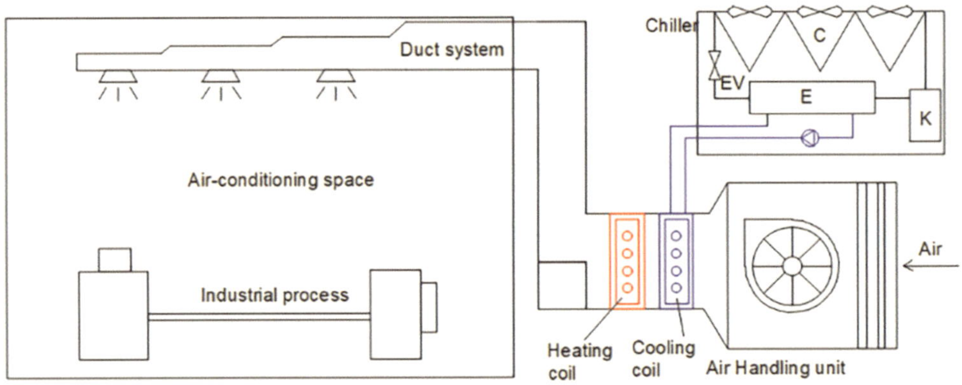

Figure 1.
Schematic diagram of an air conditioning system.

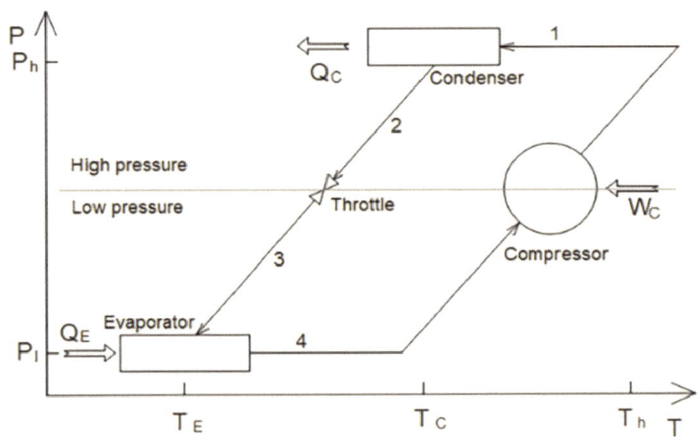

Figure 2.
Schematic diagram of an air-cooled compression chiller.

evaporator by the refrigerant at low temperature and pressure. The compressor raises the temperature and pressure of the refrigerant to a temperature higher than that of the environment so that the heat can be released through the condenser to the environment, to condense the refrigerant. The refrigerant liquid reduces its pressure in an expansion valve to bring the working fluid to the required temperature and pressure conditions at the evaporator inlet so that the cycle is repeated. Compression chillers release heat to the surrounding air if they are air-cooled, while heat is released to the water in a cooling tower if the chiller is water-cooled.

The performance of the compression chiller is defined by the coefficient of performance of the form:

$$COP_{CC} = \frac{\dot{Q}_e}{\dot{W}_c} \tag{1}$$

The operation of compression chillers depends mainly on the electrical consumption of the compressor, $\dot{W}_c$. The heat flow supplied, $\dot{Q}_{tc}$ to the power cycle to produce the electricity consumed by the compressor is given by:

$$\dot{Q}_{tc} = \frac{\dot{Q}_e}{COP_{CC} * \eta_{tc}} \tag{2}$$

where, η_{tc}, can take values between 0.25 and 0.42, depending on the thermal cycle by which electricity is produced. For example, for a cooling capacity of 10 kW and a COP of 4.5, the compressor power with Eq. (1) is 2.22 kW. With the thermal cycle efficiency of 0.25 and using Eq. (2), the heat flow required for the ammonia chiller to operate is 8.8 kW. The increase is 75%. The high electrical consumption of conventional chillers is one of the main reasons why alternative cooling systems have been sought.

1.2 Ammonia absorption chillers

Absorption chillers have been used for several decades, following the principle of absorption refrigeration proposed by Ferdinand Carre in 1859. The cycle has been updated with a recuperator as shown in **Figure 3**. The absorption chiller is composed of the condenser, expansion valve and evaporator, similar to those of the vapor compression cycle. However, the mechanical cycle compressor is replaced by a thermochemical compressor composed of the absorber, the generator, the recuperator, the pump and the solution valve (working pairs). The operation of the system is as follows: the vapor leaving the evaporator at low pressure is absorbed by the concentrated solution in the absorber. The diluted solution is sent by means of a pump to the generator where the steam is separated at high pressure by supplying heat from an external source. The concentrated solution that leaves the generator, state 8, is directed to the absorber through the recuperator and the expansion valve to repeat the cycle. The use of these refrigeration systems has been mainly with water as a refrigerant for temperatures above 0°C. Meanwhile, refrigerant ammonia has mainly been used for evaporation temperatures below 0°C. In absorption chillers for evaporation temperatures lower than 0°C, ammonia has been predominantly used as a refrigerant and water as an absorbent, ammonia-water. This can be seen in the products offered by absorption chiller manufacturers.

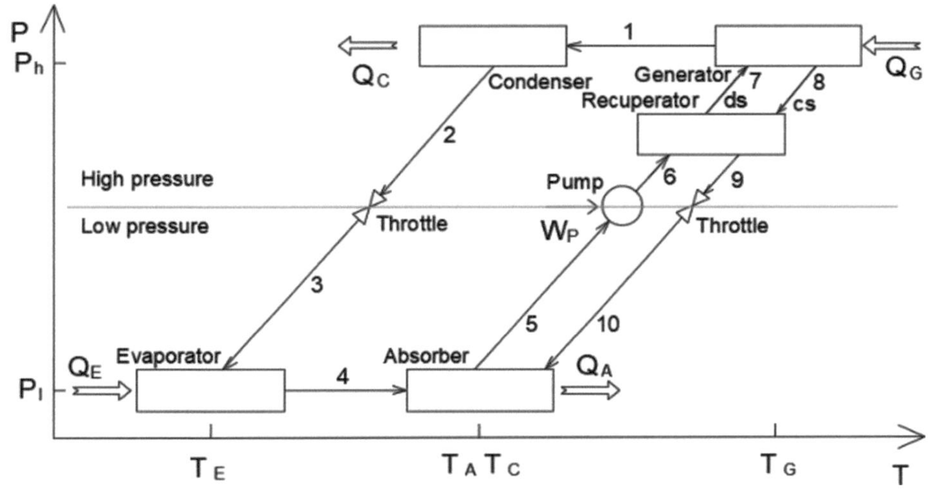

Figure 3.
Schematic diagram of the non-adiabatic absorption system.

1.2.1 Ammonia working pairs

In the research phase, absorption chillers have been found using the sodium thiocyanate ammonia solution, ammonia-sodium thiocyanate, in diffusion absorption and low-capacity refrigeration systems such as those shown by Refs. [1, 2], as well as in chillers for different applications such as shown in Refs. [3–6].

Since Infante-Ferreira presented the performance of absorption chillers in 1984, comparing several working pairs, as well as the thermophysical properties of the ammonia-lithium nitrate solution, different results of absorption chillers with ammonia-lithium nitrate have been published. The ammonia-lithium nitrate solution has been used in absorption chillers by several researchers. Such is the case of the works published by Refs. [3, 7–9] in boiling processes, or those shown by Refs. [10, 11] in microscale applications. This ammonia solution has also been used in different cooling systems, as shown by Refs. [12–16]. In this chapter, for cooling applications, absorption chillers using the ammonia-lithium nitrate solution have been divided into non-adiabatic absorption chillers, NAAC and adiabatic absorption chillers, AAC, as described below.

1.2.2 Adiabatic absorption chillers

The concept of adiabatic absorption was presented by Ref. [17], for the water-lithium bromide work pair. With the same dilution Refs. [18, 19] published the results of an air-cooled adiabatic absorption refrigeration system. A single-effect absorption refrigeration cycle driven by an ejector-adiabatic absorber was published by Ref. [20]. Meanwhile, the comparison of adiabatic and non-adiabatic absorption chillers using saline solutions was published by Ref. [21].

The configuration of an adiabatic absorption chiller is presented in **Figure 4**. The operation of this system is similar to that of non-adiabatic absorption described in **Figure 3**, with the variant that in the adiabatic system, the solution that comes out of the liquid expansion valve is subcooled in a flow heat exchanger, single-phase, prior to accessing the adiabatic absorber, in which there is no heat transfer, but only mass is transferred (vapor absorption by the solution).

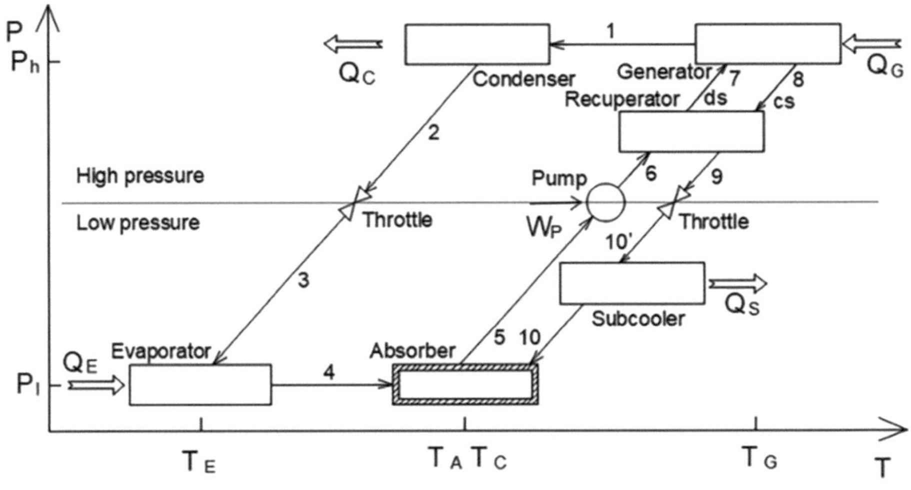

Figure 4.
Schematic diagram of an adiabatic absorption system.

1.2.3 Non-adiabatic ammonia-lithium nitrate absorption chillers

Absorption chillers, such as the one shown in **Figure 3**, using the ammonia-lithium nitrate solution, with non-adiabatic absorbers are presented in this section. Ventas et al. [22] evaluated a hybrid single-effect absorption refrigeration system with mechanical compression; Zamora et al. [23] evaluated air-cooled and water-cooled absorption chillers in the pre-industrial stage; while Ref. [15] evaluated the lithium-ammonia-nitrate solution in a prototype absorption chiller.

1.2.4 Adiabatic ammonia-lithium nitrate absorption chillers

Ammonia vapor absorption results in adiabatic absorbers were published by Refs. [14, 24–26]. Verdasco et al. [27] evaluated adiabatic absorption processes with the ammonia-water solution. While Ref. [28] evaluated the adiabatic absorption characteristics of the ammonia-sodium thiocyanate solution.

1.2.5 Advantages and disadvantages of ammonia absorption chillers

In general, among the main advantages of absorption chillers with ammonia, over conventional compression chillers with the same refrigerant, is that the former use residual and/or renewable sources for their operation [29, 30]. One of the main advantages of non-adiabatic absorption chillers over adiabatic ones that use the ammonia-lithium nitrate solution is that the former already have several prototypes and even patents, as will be described later [31–33]. Unlike adiabatic absorption chillers with the ammonia-lithium nitrate solution, at this time they have only been evaluated theoretically or experimentally. However, as already described above, the main advantages of adiabatic absorption chillers are that the configuration is simpler, that the components can be cheaper and that, like non-adiabatic absorption chillers, by using natural refrigerants and/or low Global Warming Potential, GWP, contributes to the reduction of global warming [21].

1.2.6 Ammonia-lithium nitrate absorption chiller applications

The applications of non-adiabatic absorption chillers using the ammonia-lithium nitrate solution can be, for example, air conditioning, refrigeration and freezing, as can be seen in Refs. [29, 32–34]. In the work published by Ref. [35], it can be seen that these chillers could be used to preserve medicines.

Figure 5 presents the approximate results, published by Ref. [36]. From the figure, it can be seen that for evaporation temperatures of 5°C to −30°C, absorption chillers, both those that use the ammonia-water solution such as those that use the ammonia-lithium nitrate solution, can operate with supply temperatures between 70 and 130°C. From the figure, it can also be seen that the coefficient of performance, COP with the ammonia-lithium nitrate solution with evaporation temperatures below 0°C is higher than that for the solution with water. For evaporation temperatures of 5°C, the COP is 10% lower than when the ammonia-water solution is used. Therefore, according to these observations, and the results shown by Ref. [21], it can be suggested that adiabatic absorption chillers that use the ammonia-lithium nitrate solution can be used for air conditioning, refrigeration and freezing processes. These processes are found in the conservation of food, crops, conservation of medicines, etc.

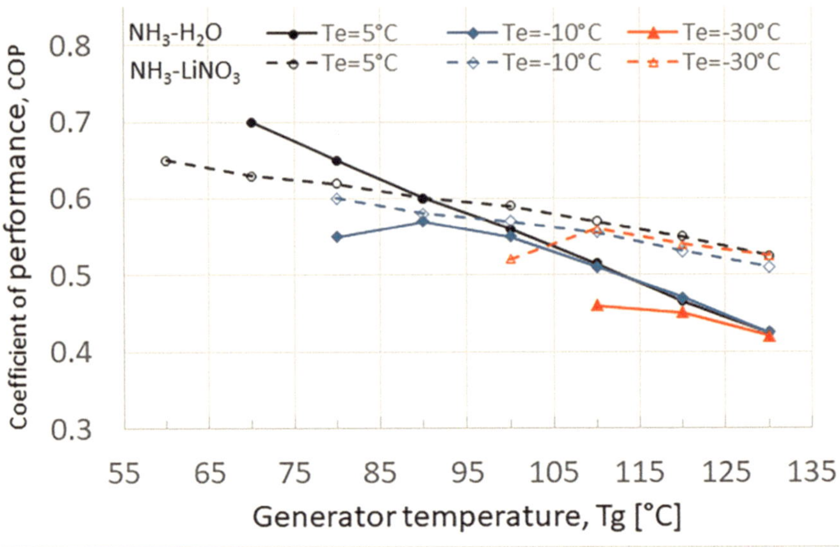

Figure 5.
Coefficient of Performance with respect to generation temperature at different evaporation temperatures for ammonia-water and ammonia-lithium nitrate solutions..

2. Analysis and mathematical modeling of absorption chillers

Absorption chiller modeling is presented in this section for adiabatic and non-diabatic refrigeration systems. The modeling presented here of the non-adiabatic absorption chiller is carried out through mass and energy balance for each component, as suggested by Refs. [21, 37], as well as the principles of heat transfer and the procedure followed for programming these systems, as described below.

2.1 Non-adiabatic absorption chiller

Since the evaporator heat flow $\dot{Q}_e$ and the evaporation temperature, T_e, is provided by the system application, then the required refrigerant flow is determined from:

$$\dot{Q}_e = \dot{m}_r(h_4 - h_3) \tag{3}$$

where h_4 and h_3 are the exit and entry enthalpies of the evaporator respectively.

The low pressure, P_l and high pressure, P_h of the system are determined by the equations of the thermodynamic properties of the refrigerant, at the temperatures of the evaporator, T_e and the condenser, T_c, respectively. The enthalpy at the condenser outlet, h_2, is the enthalpy of the saturated liquid at the condensation temperature. The enthalpy at the evaporator inlet, $h_3 = h_2$ due to the development of an isenthalpic process in the refrigerant expansion valve. The temperature of the refrigerant vapor at the inlet of the condenser is equal to the temperature of the concentrated solution at the outlet of generator T_8. This temperature will be defined by the temperature of the thermal supply source T_g, as:

$$T_8 = T_g \tag{4}$$

The temperature of the diluted solution at the outlet of the absorber, T_5, is equal to the condensation temperature, because the absorption heat, is released to the same sink as that released in the condenser. Therefore,

$$\mathrm{T}_5 = \mathrm{T}_a = T_c \tag{5}$$

Once the temperatures, T_5 and T_8, have been defined, as well as the low pressures, P_{l} and high pressures, P_{h}, the concentrations of the diluted and concentrated solution can be determined at points 5 and 8 respectively, using the equations of the thermophysical properties of the working fluid.

$$X_5 = f(P_l, T_5) \tag{6}$$

$$X_8 = f(P_h, T_8) \tag{7}$$

The change in concentration ΔX is:

$$\Delta X = X_{ds} - X_{cs} \tag{8}$$

The circulation ratio, f, can be determined as:

$$f = \frac{\dot{m}_{ds}}{\dot{m}_r} = \frac{X_r - X_{cs}}{X_{ds} - X_{cs}} \tag{9}$$

From this equation, the flow rate of the dilute solution, $\dot{m}_{ds}$, can be determined as:

$$\dot{m}_{ds} = \dot{m}_r \frac{X_r - X_{cs}}{X_{ds} - X_{cs}} \tag{10}$$

The flow of the concentrated solution, $\dot{m}_{cs}$, is determined by mass balance in the absorber, as:

$$\dot{m}_{ds} = \dot{m}_{cs} + \dot{m}_r \tag{11}$$

With the temperature and concentration of states 5 and 8, h_5 and h_8 can be determined.

The power of the pump, $\dot{W}_p$ can be determined, with the change in pressures of the form:

$$\dot{W}_p = \frac{P_h - P_l}{\eta_m} \dot{m}_{ds} v_5 \tag{12}$$

how,

$$w_p = \frac{\dot{W}_p}{\dot{m}_{ds}} \tag{13}$$

Then, the enthalpy at the pump outlet, h_6, can be determined from:

$$w_p = (h_6 - h_5) \tag{14}$$

The heat transferred in the recuperator is determined as:

$$\dot{Q}_r = \dot{m}_{ds}(h_7 - h_6) \tag{15}$$

$$\dot{Q}_r = \dot{m}_{cs}(h_8 - h_9) \tag{16}$$

While the thermal efficiency of this heat exchanger is determined as suggested by Ref. [38], by:

$$\varepsilon_r = \frac{T_7 - T_6}{T_8 - T_6} \tag{17}$$

If $C_{min} = C_h$. or by:

$$\varepsilon_r = \frac{T_8 - T_9}{T_8 - T_6} \tag{18}$$

If $C_{min} = C_c$. where C_c and C_h are the thermal capacitance of the fluids, cold or hot respectively.

$$C_c = \dot{m}_c Cp_c \tag{19}$$

$$C_h = \dot{m}_h Cp_h \tag{20}$$

Once the temperature in states T_7 and T_9 is known, the enthalpies at those points are determined. The enthalpy $h_{10} = h_9$ is due to the developing isenthalpic expansion in the solution valve. The heat flow in each of the components is determined by energy and mass balance and is as follows:

$$\dot{Q}_g = \dot{m}_r h_1 + \dot{m}_{cs} h_8 - \dot{m}_{ds} h_7 \tag{21}$$

$$\dot{Q}_a = \dot{m}_r h_4 + \dot{m}_{cs} h_{10} - \dot{m}_{ds} h_5 \tag{22}$$

$$\dot{Q}_c = \dot{m}_r (h_1 - h_2) \tag{23}$$

The operating coefficient, *COP*, is determined from:

$$COP = \frac{\dot{Q}_e}{\dot{Q}_g + \dot{W}_p} \tag{24}$$

2.2 Adiabatic absorption chiller

The mathematical analysis and modeling of adiabatic absorption chillers are carried out as indicated below.

First, the independent variables are assigned to each key state: $T_1 = T_8 = T_g$, $T_{10} = T_2 = T_c, T_4 = T_e$. With this and with the equations of the thermodynamic properties of the refrigerant, the operating pressures, low pressure and high pressure, as well as the enthalpies of the four states of the refrigerant in the refrigerant circuit, are determined.

With the required cooling capacity of the system, the refrigerant flow is determined with Eq. (3). The concentration of the concentrated solution can be determined by Eq. (8). While the enthalpy in states 8 and 10, is determined by:

$$h_8 = f(T_8, X_{cs}) \tag{25}$$

$$h_{10} = f(T_{10}, X_{cs}) \tag{26}$$

From here, the missing variables in states 5–10′ are determined by the system of Eqs. (9)–(20), in addition to Eqs. (27) and (28) obtained from the energy and mass balance of the adiabatic absorber.

$$\dot{m}_r X_r + \dot{m}_{cs} X_{cs} = \dot{m}_{ds} X_{ds} \tag{27}$$

$$\dot{m}_r h_4 + \dot{m}_{cs} h_{10} = \dot{m}_{ds} h_5 \tag{28}$$

The heat flow released by the subcooler is determined by:

$$\dot{Q}_s = \dot{m}_{cs}(h_{10'} - h_{10}) \tag{29}$$

3. Performance of absorption chillers with ammonia-lithium nitrate

The model presented in Section 2 can be programmed in computer software to simulate the behavior of adiabatic and non-adiabatic absorption chillers at different operating conditions. In this section, the performance results of adiabatic and non-adiabatic absorption chillers using the ammonia-lithium nitrate solution are presented. The results were obtained using the Engineering Equation Solver, EES software [39]. The thermodynamic properties of the solution were determined using the correlations of Refs. [40, 41] and were programmed in the EES software. The cooling power, and the efficiency of the recuperator and the pump were, respectively, 10 kW, 0.8, 0.85. While the simulated temperature values are presented in **Table 1**.

Figure 6 shows the behavior of the heat flow in the generator, with respect to the generation temperature at different evaporation temperatures, for two condensation temperatures and for adiabatic and non-adiabatic chillers. In the figure, in general terms, it can be seen that the heat flow required for an adiabatic chiller is up to 50% greater than that needed in a non-adiabatic chiller at 90°C or 28% at 110°C. This may be due, as explained in Ref. [21], that the concentration change is smaller in the absorber for an adiabatic chiller because only the sensible heat is extracted in the subcooler. Consequently, a greater flow of solution is required to circulate in the absorber and in the generator to separate the amount of refrigerant at the required cooling capacity. In the same figure, it can be seen that at a higher evaporation temperature and a lower condensation temperature, for example at 0°C and 28°C respectively, the supplied heat flux increases only 22% for the adiabatic system compared to the non-adiabatic system.

Figure 7 shows the behavior of the circulation ratio, with respect to the generation temperature at different evaporation temperatures, for two condensation temperatures and for adiabatic and non-adiabatic chillers. Like the heat flow previously discussed in **Figure 6**, it can be seen that the circulation ratio for the adiabatic chiller is 78% greater compared to that required by the non-adiabatic chiller at 90°C or 74% at 110°C.

Parameter	Value
Evaporator temperature, T_e, °C	−10, −5, 0
Condenser temperature, T_c, °C	28, 31, 34
Generator temperature, T_g, °C	90–110

Table 1.
Operating conditions for simulating.

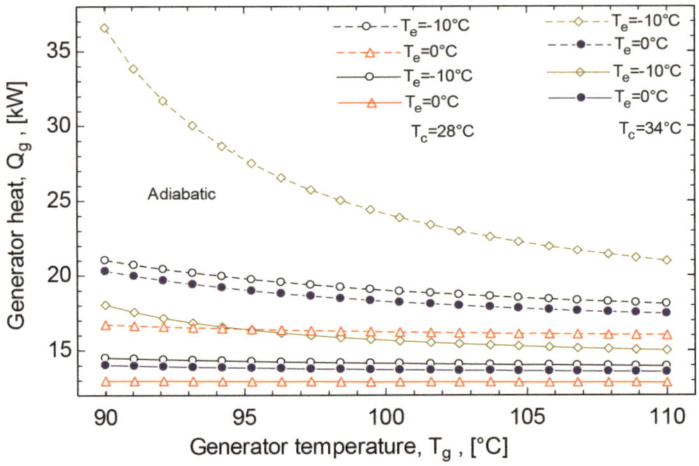

Figure 6.
Heat flow in the generator with respect to the generation temperature, for different evaporation temperatures at two condensation temperatures and for adiabatic (dashed lines) and non-adiabatic (continuous lines) chillers.

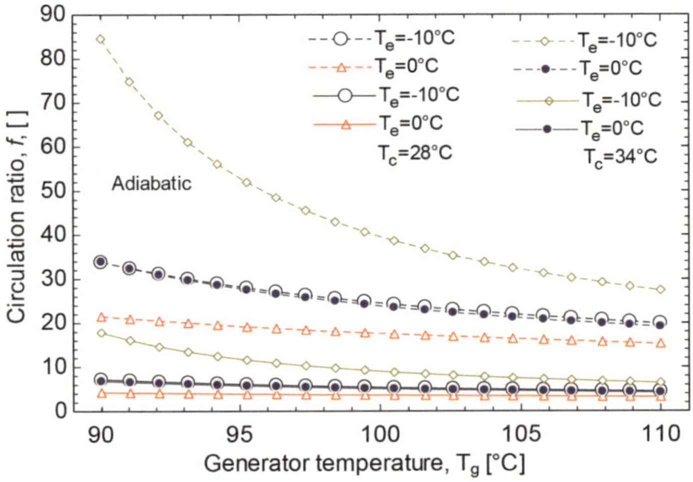

Figure 7.
Circulation ratio with respect to the generation temperature, for different evaporation temperatures at two condensation temperatures.

Figure 8 shows the behavior of the coefficient of performance, with respect to the generation temperature at different evaporation temperatures, for two condensation temperatures and for adiabatic and non-adiabatic chillers. From the figure, it can be seen that lower generation temperature, the *COP* is lower. Likewise, it is observed that the *COP* of the adiabatic chiller is lower than that of the non-adiabatic chiller. However, for example, at 95.2°C, it can also be noted that at a higher evaporation temperature and a lower condensation temperature, for example at 0°C and 28°C, respectively, the *COP* of the adiabatic chiller decreases only approximately 20%, compared to that of the non-adiabatic chiller. From the results presented in this section, it can be seen that although the heat flow and circulation ratio are higher and the cop is lower for absorption chillers, when the condensation temperature is lower

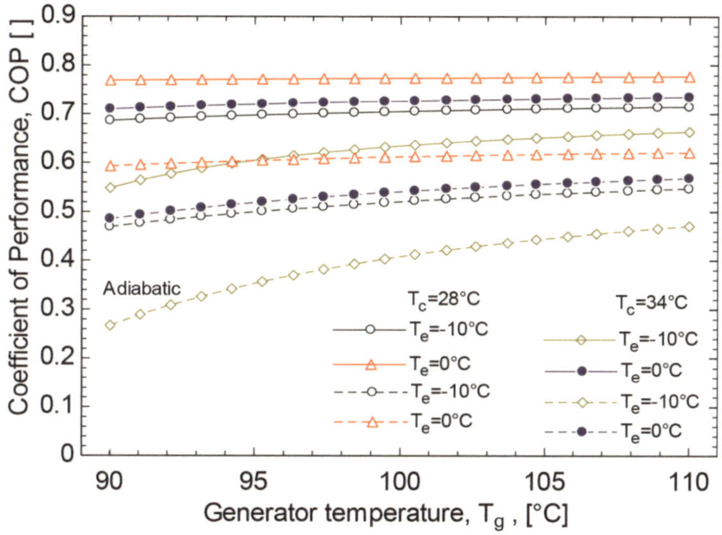

Figure 8.
Coefficient of performance with respect to the generation temperature, for different evaporation temperatures at two condensation temperatures.

and the evaporation and generation temperatures are higher, these design parameters are closer to those of non-adiabatic absorption chillers. The above suggests further studies in these cycles.

4. Future research on adiabatic ammonia-lithium nitrate absorption chillers

According to the results published by Refs. [21, 29, 35, 42], the results presented in this chapter and the trend in research on refrigeration systems, alternative energy sources and the need to reduce global warming, among others, some future lines of research on adiabatic absorption chillers using the ammonia-lithium nitrate solution may be:

- Evaluation of adiabatic absorption chillers using nanofluids to improve heat transfer.
- Evaluation of double-effect and triple-effect adiabatic absorption chillers to reduce supply energy.
- Evaluation of adiabatic absorption chillers integrated into polygeneration systems, to improve primary energy efficiency.

5. Conclusions

From the information presented in this chapter, the following conclusions can be found.

Although the performance coefficient of traditional ammonia compression chillers is greater than 4, when the cycle and the thermal source for the production of electricity consumed by the cycle compressor are taken into account, the energy required for the operation of the cooling system cooling is up to 75% greater than the required cooling capacity.

On the other hand, although the principle of adiabatic absorption in absorption chillers has been proposed for more than three decades, there are very few published works on adiabatic absorption chillers with ammonia as a refrigerant. This suggests that more research is required in this regard, in order to explore the benefits of this cycle.

From the results presented in this section, it can be seen that although the heat flow and circulation ratio are higher and the cop is lower for absorption chillers, when the condensation temperature is lower and the evaporation and generation temperatures are higher, these design parameters are closer to those of non-adiabatic absorption chillers. The above suggests further studies in these cycles.

Acknowledgements

The authors thank the Instituto Politécnico Nacional for funding through the research project: SIP20240435.

Conflict of interest

The authors declare no conflict of interest.

Nomenclature

AAC	adiabatic absorption chiller
COP	coefficient of performance
Cp	specific heat (kJ/kg K)
f	circulation ratio
h	enthalpy (kJ/kg)
NAAC	non-adiabatic absorption chiller
$\dot{m}$	mass flow rate (kg/s)
$\dot{Q}$	heat flow (W)
T	temperature (K)
v	specific volume (m^3/kg)
$\dot{W}$	power (W)
x	concentration

Subscripts

1,2,3… 10	state in the cycle
a	absorber
ac	absorption chiller
c	condenser
cc	compression chiller

cs	concentrated solution
ds	diluted solution
e	evaporator
g	generator
p	pump
r	refrigerant
s	subcooler
tc	thermal cycle

Greek symbols

ε,η	efficiency

Author details

Alejandro Zacarías[1*], Guerlin Romage[1], Cuauhtémoc Jiménez[1], José de Jesús Rubio[1], Ignacio Carvajal[2] and María Venegas[3]

1 Instituto Politécnico Nacional, ESIME Azcapotzalco, Mexico City, Mexico

2 Instituto Politécnico Nacional, UPALM, Mexico City, Mexico

3 Universidad Carlos III de Madrid, Ingeniería Térmica y de Fluidos, Madrid, Spain

*Address all correspondence to: azacarias@ipn.mx

References

[1] Acuña A, Velázquez N, Cerezo J. Energy analysis of a diffusion absorption cooling system using lithium nitrate, sodium thiocyanate and water as absorbent substances and ammonia as the refrigerant. Applied Thermal Engineering. 2013;**51**:1273-1281

[2] Mendoza LC, Ayou DS, Navarro-Esbrí J, et al. Small capacity absorption systems for cooling and power with a scroll expander and ammonia based working fluids. Applied Thermal Engineering. 2014;**72**:258-265

[3] Cai D, Jiang J, He G, et al. Experimental evaluation on thermal performance of an air-cooled absorption refrigeration cycle with NH3-LiNO3 and NH3-NaSCN refrigerant solutions. Energy Conversion and Management. 2016;**120**:32-43

[4] Garousi Farshi L, Asadi S. Ammonia lithium nitrate and ammonia sodium thiocyanate double effect absorption refrigeration systems: Thermodynamic analysis. Applied Thermal Engineering. 2018;**138**:374-385

[5] Zhou S, He G, Li Y, et al. Comprehensive experimental evaluation of an exhaust-heat-driven absorption refrigeration cycle system using NH3-NaSCN as working pair. International Journal of Refrigeration. 2021;**126**:168-180

[6] Ngock GRH, Tamba JG, Djanna F, et al. Analysis and improvement of the efficiency of NH3-NaSCN single effect absorption cooling system. Heliyon. 2022;**8**:1-13. DOI: 10.1016/j.heliyon.2022.e11635

[7] Zacarías A, Ventas R, Venegas M, et al. Boiling heat transfer and pressure drop of ammonia-lithium nitrate solution in a plate generator. International Journal of Heat and Mass Transfer. 2010;**53**:4768-4779. DOI: 10.1016/j.ijheatmasstransfer.2010.06.015

[8] Venegas M, Zacarías A, Vereda C, et al. Subcooled and saturated boiling of ammonia-lithium nitrate solution in a plate-type generator for absorption machines. International Journal of Heat and Mass Transfer. 2012;**55**:4914-4922. DOI: 10.1016/j.ijheatmasstransfer.2012.04.061

[9] Taboas F, Bourouis M, Valles M. Boiling heat transfer and pressure drop of $NH_3/LiNO_3$ and $NH_3/(LiNO_3+H_2O)$ in a plate heat exchanger. International Journal of Thermal Sciences. 2016;**105**:182-194

[10] Amaris C, Bourouis M, Vallès M. Passive intensification of the ammonia absorption process with $NH_3/LiNO_3$ using carbon nanotubes and advanced surfaces in a tubular bubble absorber. Energy. 2014;**68**:519-528

[11] Liang X, He G, Wang J, et al. Experimental study on absorption characteristics of a falling film absorber with micro-scale $NH_3/LiNO_3$ liquid film. Applied Thermal Engineering. 2022;**200**:1-18. DOI: 10.1016/j.applthermaleng.2021.117719

[12] Rivera W, Moreno-Quintanar G, Rivera CO, et al. Evaluation of a solar intermittent refrigeration system for ice production operating with ammonia/lithium nitrate. Solar Energy. 2011;**85**:38-45

[13] Hernández-Magallanes JA, Domínguez-Inzunza LA, Gutiérrez-Urueta G, et al. Experimental assessment of an absorption cooling system operating with the ammonia/lithium nitrate mixture. Energy. 2014;**78**:685-692

[14] Ventas R, Lecuona A, Vereda C, et al. Performance analysis of an absorption double-effect cycle for power and cold generation using ammonia/lithium nitrate. Applied Thermal Engineering. 2017;**115**:256-266

[15] Altamirano A, Le Pierrès N, Stutz B, et al. Performance characterization methods for absorption chillers applied to an NH_3-$LiNO_3$ single-stage prototype. Applied Thermal Engineering;**185**:1-13. Epub ahead of print 25 February 2021. DOI: 10.1016/j.applthermaleng.2020.116435

[16] Colorado D, Rivera W, Conde-Gutiérrez RA, et al. Energy and exergy analysis of an experimental NH_3-$LiNO_3$ air-conditioning absorption system. International Journal of Refrigeration. 2024;**165**:393-405

[17] Ryan William A. Water absorption in an adiabatic spray of aqueous lithium bromide solution. In: AES-Vol. 31, International Absorption Heat Pump Conference, ASME; 1993. pp. 155-162

[18] Wang L, Chen GM, Wang Q, et al. Thermodynamic performance analysis of gas-fired air-cooled adiabatic absorption refrigeration systems. Applied Thermal Engineering. 2007;**27**:1642-1652

[19] Osta-Omar SM, Micallef C. Mathematical model of a lithium-bromide/water absorption refrigeration system equipped with an adiabatic absorber. Computation. 2016;**4**:1-16. DOI: 10.3390/computation4040044

[20] Vereda C, Ventas R, Lecuona A, et al. Single-effect absorption refrigeration cycle boosted with an ejector-adiabatic absorber using a single solution pump. International Journal of Refrigeration. 2014;**38**:22-29

[21] Zacarías A, Quiroz JA, Gutiérrez-Urueta GL, et al. Comparison between adiabatic and non-adiabatic absorption chillers using ammonia-lithium nitrate and water-lithium bromide solutions. Heat Transfer Research. 2020;**51**:609-621

[22] Ventas R, Lecuona A, Zacarías A, et al. Ammonia-lithium nitrate absorption chiller with an integrated low-pressure compression booster cycle for low driving temperatures. Applied Thermal Engineering. 2010;**30**:1351-1359. DOI: 10.1016/j.applthermaleng.2010.02.022

[23] Zamora M, Bourouis M, Coronas A, et al. Pre-industrial development and experimental characterization of new air-cooled and water-cooled ammonia/lithium nitrate absorption chillers. International Journal of Refrigeration. 2014;**45**:189-197

[24] Zacarías A, Venegas M, Ventas R, et al. Experimental assessment of ammonia adiabatic absorption into ammonia-lithium nitrate solution using a flat fan nozzle. Applied Thermal Engineering. 2011;**31**:3569-3579. DOI: 10.1016/j.applthermaleng.2011.07.019

[25] Zacarías A, Venegas M, Lecuona A, et al. Experimental evaluation of ammonia adiabatic absorption into ammonia-lithium nitrate solution using a fog jet nozzle. Applied Thermal Engineering. 2013;**50**:781-790. DOI: 10.1016/j.applthermaleng.2012.07.006

[26] Zacarías A, Venegas M, Lecuona A, et al. Experimental assessment of vapour adiabatic absorption into solution droplets using a full cone nozzle. Experimental Thermal and Fluid Science. 2015;**68**:228-238. DOI: 10.1016/j.expthermflusci.2015.05.001

[27] Berdasco M, Coronas A, Vallès M. Study of the adiabatic absorption process in polymeric hollow fiber membranes for ammonia/water absorption refrigeration

systems. Applied Thermal Engineering. 2018;**137**:594-607

[28] Liang X, He G, Zhou S, et al. Absorption characteristics of NH_3/NASCN working pair in an adiabatic absorber with structured packing. Applied Thermal Engineering. 2021;**185**: 1-13. DOI: 10.1016/j.applthermaleng.2020.116325

[29] Nikbakhti R, Wang X, Hussein AK, et al. Absorption cooling systems – Review of various techniques for energy performance enhancement. Alexandria Engineering Journal. 2020;**59**:707-738

[30] Redacción, Refrigeración con Amoníaco [on line]. Mundo HVAC&R. 2014. Available from: https://www.mundohvacr.com/2009/03/refrigeracion-con-amoniaco/ [Accessed: January 10, 2025]

[31] Bourouis M, Salcedo AC, Raquera JMV, Garcia MZ. Enfriadora de agua por absorción tipo aire-agua o agua-agua de amoniaco y nitrato de litio. WO2011039397. 2011

[32] Hernandez JI, Rivera W, Franco G, et al. Sistema de enfriamiento solar vertical operando con la mezcla nitrato de litio-amoniaco. MX 365474 B. 2013

[33] Gutierrez JIH, Franco WRG, Brown RB, Valladares OG, Espinoza VHG, Magallanes JAH. Sistema de enfriamiento solar horizontal operando con la mezcla nitrato de litio-amoniaco. 365473. 2019

[34] Wu W, Wang B, Shi W, et al. An overview of ammonia-based absorption chillers and heat pumps. Renewable and Sustainable Energy Reviews. 2014;**31**: 681-707

[35] Rivas D, García AI, Barrera EE, Sánchez JG, Romage G, et al. Modelación y simulación de un refrigerador solar para medicamentos mediante refrigeración por absorción. In: XII Congreso Ibérico y X Congreso Iberoamericano de Ciencias y Técnicas del Frío, Elche Spain. 2024. DOI: 10.21134/hpy1ch59

[36] Infante-Ferreira CA. Vertical tubular absorbers for ammonia-salt absorption refrigeration. [PhD Thesis]. Delft University of Technology; 1985

[37] Herold Keith E, Reinhard R, Klein SA. Absorption Chillers and Heat Pumps. 2nd ed. Columbus, OH, USA: CRC Press; 2016

[38] Incropera FP, DeWitt DP, Fundamentals of Heat and Mass Transfer. 4th ed. John Wiley & Sons Inc; 1996

[39] Klein SA. Engineering Equation Solver, 1992-2015, f-Chart Software

[40] Libotean S, Salavera D, Valles M, Esteve X, Coronas A. Vapor-liquid equilibrium of ammonia + lithium nitrate + water and ammonia + lithium nitrate solutions from (293.15 to 353.15). Journal of Chemical & Engineering Data. 2007;**52**:1050-1055

[41] Libotean S, Martín A, Salavera D, Valles M, Esteve X, Coronas A. Densities, viscosities and heat capacities of ammonia-lithium nitrate and ammonia-lithium nitrate-water solutions between (293.15 and 353.15) K. Journal of Chemical & Engineering Data. 2008;**53**:2383-2388

[42] Ayou DS, Coronas A. New developments and progress in absorption chillers for solar cooling applications. Applied Sciences (Switzerland). 2020;**10**:1-35. DOI: 10.3390/app10124073

Chapter 4

Low-Temperature Sintering Technologies in Power Electronics: Materials, Process, and Advanced Packaging of SiC WBG Semiconductors

Haidong Yan, Wanli Li, Yaqin Liao and Chaohui Liu

Abstract

SiC power devices represent a breakthrough in wide bandgap semiconductor technology, offering superior performance through high current density, extreme temperature operation, and fast switching capabilities that enable transformative improvements in power electronics, yet their full potential is constrained by the limitations of traditional soldering techniques in meeting stringent high-temperature stability and thermal management requirements. This chapter comprehensively examines low-temperature sintering technology as a transformative solution for SiC interconnects, showcasing material innovations including advanced Ag sintering formulations with nano/micro hybrid particles achieving robust connections at unprecedented low temperatures through particle engineering breakthroughs, alongside pioneering Cu sintering methods utilizing formic acid reduction and nanoscale surface treatments to provide cost-effective alternatives with exceptional mechanical strength exceeding 130 MPa. The implementation of this technology offers significant advantages across advanced packaging architectures, including DTS, Cu Clip, and double-sided cooling systems, as well as emerging embedded configurations. These developments firmly establish low-temperature sintering as an essential enabling technology for next-generation power electronics packaging solutions.

Keywords: die-attach, sintered silver/copper, wide bandgap semiconductors, power module, advanced packaging

1. Introduction

As a novel type of wide-bandgap (WBG) semiconductor, silicon carbide (SiC) power devices boast a current-carrying capacity of up to 600 A/mm^2, potential for extreme high-temperature service above 250°C, and high-frequency switching performance.

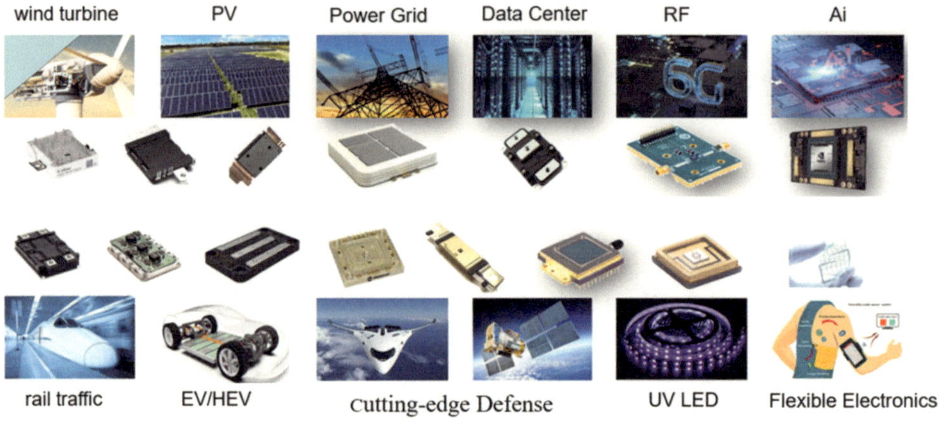

Figure 1.
Potential application scenarios for low-temperature sintering technologies [2–6].

These characteristics are revolutionary for enhancing the power density and conversion efficiency of power electronic systems, and consequently reducing system costs. SiC power devices have attracted considerable attention in fields, such as new energy electric vehicles, high-voltage direct current transmission, rail transportation, and cutting-edge defense equipment, and they have gradually been widely applied in new energy electric vehicles. However, there are significant differences in electrical, thermal, mechanical, and other performance aspects between SiC power devices and traditional Si-based power devices. Traditional soft soldering alloys can no longer meet the high-performance requirements of SiC devices for die bonding materials in terms of high-temperature resistance, efficient heat dissipation, process compatibility, and so on.

Low-temperature sintering silver (Ag) or copper (Cu) materials are novel high-performance die interconnect materials capable of forming a die attachment with high-temperature resistance, high thermal conductivity, and low resistivity under certain auxiliary pressure or even without pressure at low temperatures [1]. Compared to conventional soft soldering alloys, die attachment formed by low-temperature silver/copper sintering technology not only enhances the high-temperature service reliability and reduces the thermal resistance of SiC power packaging, but also eliminates remelting issues during multilevel SiC power packaging. Thus, low-temperature sintering technologies demonstrate broad application prospects in SiC power packaging and even high thermal dissipation microelectronic packaging fields. Potential application scenarios for low-temperature sintering technologies are shown in **Figure 1**. This paper provides a concise overview of the sintering mechanisms, processing methodologies, and representative applications in state-of-the-art wide-bandgap semiconductor packaging for low-temperature silver/copper sintering materials.

2. Low-temperature sintering materials

2.1 Low-temperature sintering materials vs. traditional soft soldering alloys

Soldering remains the most widely used interconnect technology in electronics manufacturing, primarily due to its simplicity and suitability for mass production

(**Figure 2**) [9]. The standard method for applying solder paste involves a stencil-printing process, where a viscous solder paste is spread over a thin metal stencil and deposited onto the substrate [10]. This solder paste is composed of tiny metal spheres and a binder material that includes flux. Once the solder paste is applied, components are precisely positioned using an automated pick-and-place machine onto the printed solder deposits [11]. The assembled substrates are then passed through a reflow oven, which is divided into multiple temperature zones to achieve an optimal soldering profile for the paste [12–14]. The soldering profile typically consists of three key stages: preheat/soak, reflow, and cool down. The schematic representation of the conventional soft soldering process is illustrated in **Figure 3**.

For nearly 20 years, low-temperature sintering technologies have gained recognition as a dependable alternative to traditional soldering, particularly for high-temperature applications. Sintering is a mass transport process driven by the reduction of surface energy and can be divided into three primary stages: the initial stage, where necks form between contacting particles [15]; the intermediate stage, marked by the creation of interconnected pores and the initiation of grain growth; and the final stage, characterized by the development of isolated pores, which eliminates obstacles to further grain growth. As an advanced packaging solution, sintering enables the creation of interconnects at relatively low temperatures (<300°C) while achieving a near-bulk-like microstructure with finely distributed micro- and nanopores. Consequently, silver/copper-sintered interconnects exhibit thermal, electrical, and mechanical properties that surpass those of conventional solder materials, along with the ability to operate at high temperatures (>200°C). **Figure 4** illustrates the joint formation processes and comparative characteristics between low-temperature sintering and conventional soft soldering technologies.

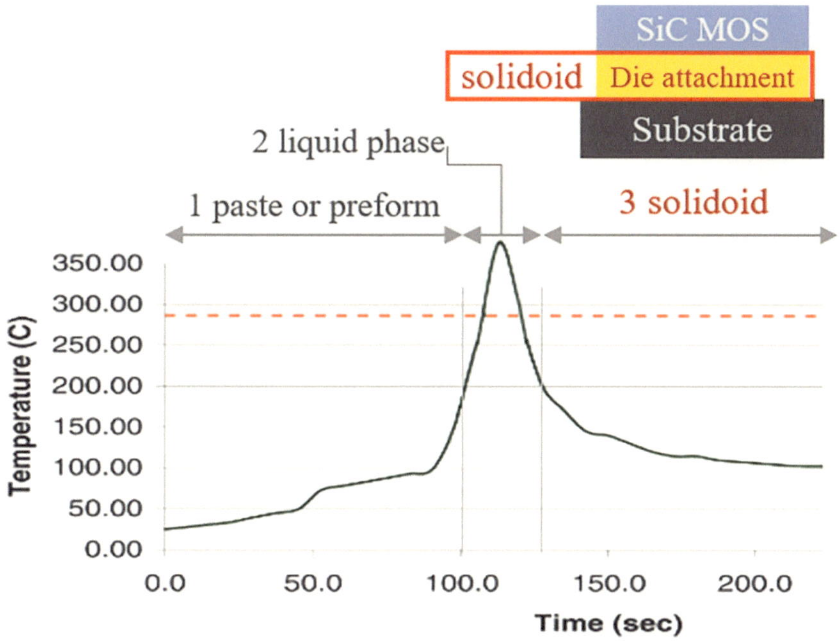

Figure 2.
Schematic diagram of traditional soft soldering process [7, 8].

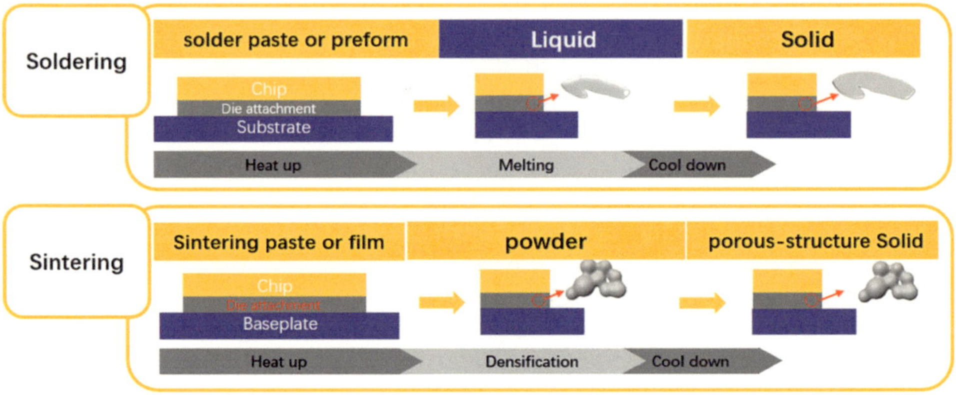

Figure 3.
The difference between silver sintering technology and traditional soft soldering technology [7, 8].

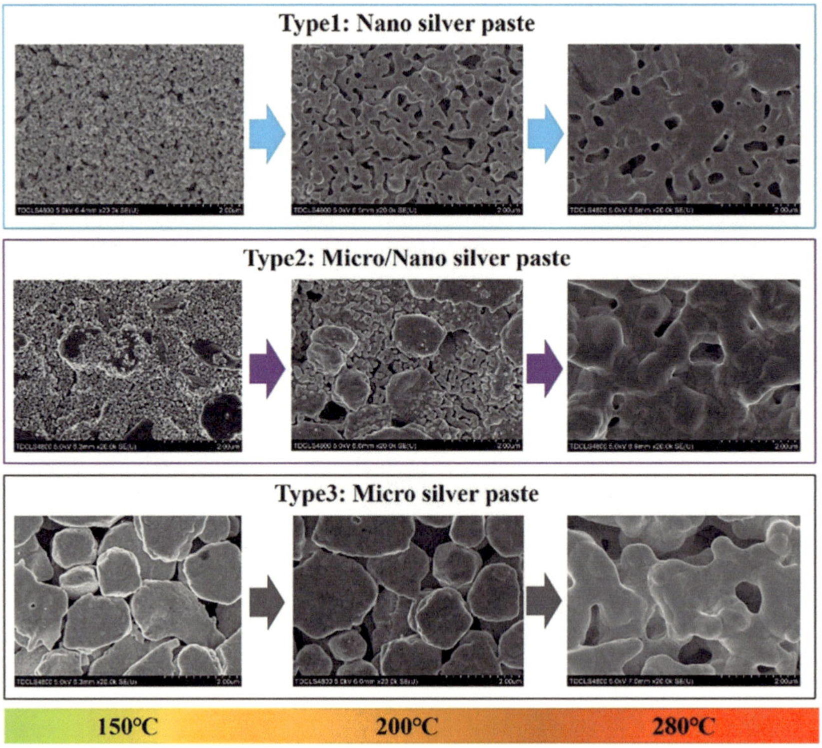

Figure 4.
The sintered microstructures of nano-, micro-nano hybrid, and micro silver pastes at 150°C, 200°C, and 280@0 MPa.

2.2 Low-temperature sintering Ag

The sintered joints formed by nano-silver paste exhibit a porous structure, with porosity decreasing as energy input increases [16]. Research demonstrates that silver particle size and morphology significantly affect joint performance: nanoparticles

(<100 nm) reduce sintering temperature and enhance driving force through size effects, while micro/nano hybrid pastes achieve higher shear strength as nanoparticles act as "bridges" [17–19]. Notably, flake-shaped silver particles enable efficient sintering at 200°C due to their high specific surface area and dense dislocations on (111) crystal planes [20]. Recent studies have further achieved pressureless sintering at ultralow 160°C through optimized solvent systems [21], where the sintering mechanism involves nanoscale necking and amorphous Ag-O structures forming at particle edges as connecting bridges (**Figure 4**) [22]. This structure maintains high strength after 1000 thermal cycles, demonstrating exceptional reliability.

2.3 Low-temperature sintering Cu

Although the low-temperature sintering silver material exhibits excellent electrical and thermal conductivity, its susceptibility to electromigration and high cost severely limit its large-scale application in power electronics. In contrast, the Cu sintering material demonstrates significant comprehensive advantages: not only does it possess comparable electrical and thermal conductivity to silver at a much lower cost, but it also exhibits superior resistance to electromigration (**Figure 5**) [23].

In recent years, breakthroughs have been made in low-temperature Cu sintering technology: by optimizing the morphology and size of Cu particles, the use of 5 nm ultrafine Cu powder enables sintering at 250°C, while formic acid vapor-assisted processes can further reduce the sintering temperature to 230°C [24]. To address the critical challenge of copper oxidation, researchers have developed three innovative solutions: formic acid vapor reduction, ascorbic acid treatment, and self-decomposing Cu salt technology [25]. Notably, sintering Cu pastes developed through surface modification and micro-nano hybrid technology can achieve high-strength joints with >130 MPa shear strength at 230°C@10 MPa, significantly outperforming traditional Sn-based soldering alloys [26]. Our team's research indicates that Cu sintering in formic acid atmosphere requires lower sintering temperature. The microstructures of sintering Cu under nitrogen and formic acid atmospheres at different temperatures are illustrated in **Figure 6**. Low-temperature Cu sintering technology is currently undergoing rapid development, and it is believed that this technology holds great promise for providing SiC power device packaging with a new interconnection solution that combines both low cost and high reliability [27–30].

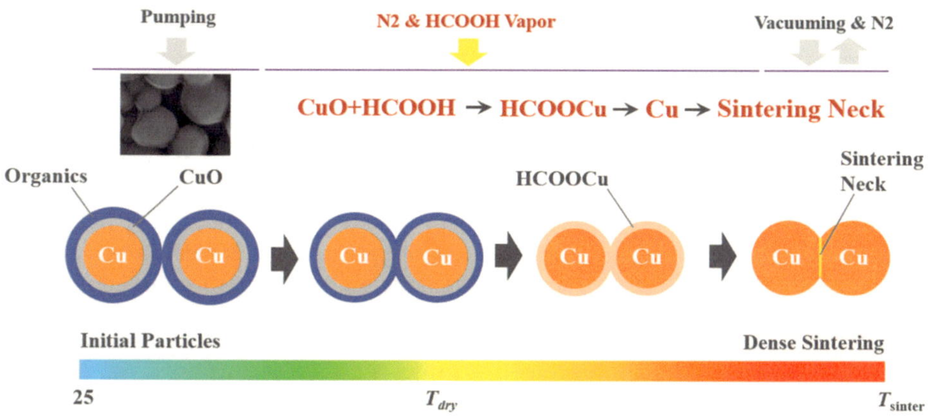

Figure 5.
Contribution of formic acid-assisted low-temperature copper sintering.

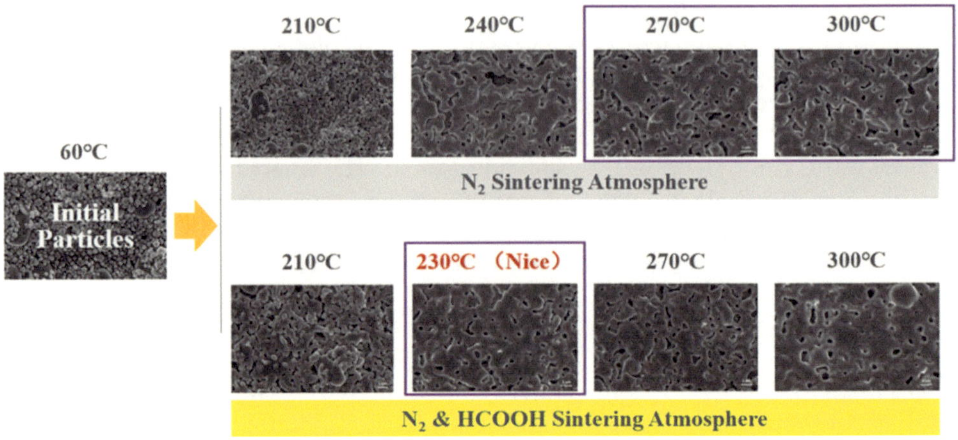

Figure 6.
The microstructures of sintering Cu under nitrogen and formic acid atmospheres at different temperatures.

2.4 Market trends and partial suppliers

In 2023, the global market share of low-temperature silver solder paste was approximately USD 360 million, with applications spanning automotive electronics, photovoltaics, ultraviolet light-emitting diodes (UV LEDs), flexible electronics, radio frequency (RF), and other fields, among which new energy vehicles represent a rigid demand [31]. The main drivers of growth in the market share of low-temperature silver solder paste over the next decade are: (1) The significant cost reduction of SiC power devices, which are penetrating into more high-power silicon-based power electronic systems or equipment, has paved the way for gradually replacing some silicon-based devices. It is estimated that within 5 to 10 years, the cost of these devices will decrease by more than 80%, approaching that of comparable silicon-based devices; (2) Continuous advancements in ultra-low-temperature or pressureless sintering technology, packaging integration technology, and other aspects related to low-temperature silver solder paste [32]. Low-temperature silver solder paste is expected to replace traditional solder alloys and conductive silver pastes in microelectronic packaging for high-power density applications (such as high-power RF devices for radar and UV LEDs). The market space in the field of microelectronic packaging is immense; (3) In the fields of flexible wearable electronic devices and packaging for ultra-high heat flux density artificial intelligence Graphics Processing Unit (AI GPU) devices, high-performance low-temperature silver solder paste also has a certain market space, but it depends on the deep integration of low-temperature silver solder paste with packaging integration methods [33–35]. In summary, the market share of pressure sintering for low-temperature silver solder paste is mainly increasing in high-reliability fields, such as new energy vehicles, rail transit, high-voltage high-capacity devices, and cutting-edge defense equipment. The pressureless sintering market is mainly valued in the field of high heat flux density microelectronic packaging, partially replacing solder alloys and conductive silver pastes. It is estimated that the global share of low-temperature silver solder paste will exceed USD 1.3 billion in 2032. The trend in the market share of low-temperature silver solder paste from 2023 to 2032 is shown in **Figure 7**.

The main suppliers of low-temperature silver paste in the world include MacDermid Alpha and Indium Corporation in the United States, and Heraeus and Henkel in Germany.

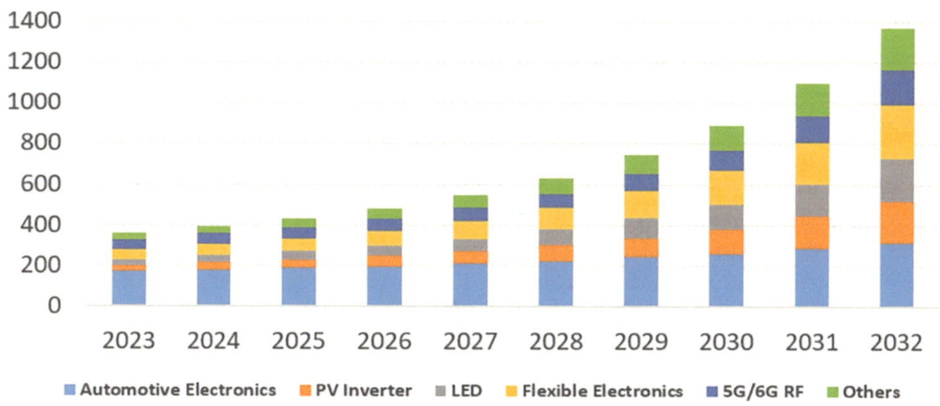

Figure 7.
Market share trend of low-temperature silver materials.

Among them, Alpha is the first to launch conductive silver glue for electronic packaging, and after 2000, it developed low-temperature sintering silver paste and used it in automotive power modules, which laid the foundation for the silver paste industry; Indium developed silver-containing thermal interface materials to improve heat dissipation performance, and the subsequent sintering silver paste was also widely used in power device interconnection; Heraeus has become the world's largest supplier of sintering silver paste; Henkel mainly innovates silver paste for flexible electronics and low-temperature curing silver paste. The silver paste of Japanese companies focuses more on the precision of materials, such as KYOCERA's silver thick film paste that is resistant to high temperatures (>800°C), and it has developed silver-copper composite paste to reduce costs and maintain high conductivity; Hitachi focuses on silver nano-ink; DOWA, as a supplier of high-purity silver powder (>99.95%), provides raw materials for silver paste companies. In recent years, China has witnessed the emergence of numerous start-ups specializing in low-temperature sintering silver paste technology, predominantly founded upon scientific innovations from universities and research institutions. Representative examples include *Xinyuan New Materials*, *QLsemi*, *Hanyuan*, *ERACHIP INNOVATIONS*, and *NAYU*, among others. The low-temperature silver paste developed by these domestic start-ups is still in the enterprise verification stage. Due to the weak risk tolerance of start-ups and the high-reliability requirements of automobile companies for low-temperature silver paste, its application in vehicles is not yet mature and requires more time for verification.

3. Low-temperature sintering mechanism and process effect factors

3.1 Low-temperature sintering mechanism

The low-temperature sintering connection of particles is a process of obtaining a strong bond and densification by contacting, interdiffusion, deformation, and rearrangement of particles through external heat and auxiliary pressure; that is, under the action of the sintering driving force, due to their own size effect, the atoms of the particles directly contact and diffuse with each other to form a tight connection [36]. The meaning of the size effect is that nano- and microparticles have higher surface energy due to their smaller size and are more likely to undergo material migration at lower temperatures to promote the sintering process [37].

At the same time, due to the high melting point characteristics of silver and copper materials, the effect of "low temperature connection and high temperature service" can be achieved. The sintering stage of low-temperature sintering connection can be divided into three stages: initial sintering, mid-sintering, and late sintering. In the initial stage of sintering, the particles move under the action of the sintering driving force until they contact each other and begin to form a sintering neck. At this time, the degree of densification is low. In the mid-sintering stage, the sintering neck grows further, and the sintering neck shrinks significantly to form cylindrical pores. The shape of the particles gradually changes from spherical to polyhedral, and the grains grow and the degree of densification is significant.

In the later stage of sintering, the diameter of the sintering neck reaches the maximum, the pores gradually change from a connected state to a closed state, the pores continue to decrease and tend to be spherical, and the degree of densification reaches the highest level. **Figure 1** uses a two-equal-diameter particle model to describe the diffusion mechanism of atoms during sintering, where the red arrow curve represents the possible diffusion path, and R and X are the particle radius and sintering neck, respectively. **Figure 8** shows several common sintering mechanisms, which are surface diffusion, vapor transmission, surface lattice diffusion, grain boundary lattice diffusion, grain boundary diffusion, and plastic deformation.

The sintering neck is the product of the convergence of atomic diffusion paths under all the above sintering mechanisms. The growth of the sintering neck can be further described by the empirical Formula (1), where R is the particle radius, X is the sintering neck radius, t is the sintering time, H is the characteristic coefficient related to the sintering mechanism, and m and n are constants related to the sintering diffusion mechanism [38].

$$\left(\frac{X}{R}\right)^m = \frac{H}{R^n}t \tag{1}$$

Surface diffusion, grain boundary diffusion, and lattice diffusion are the main diffusion mechanisms in the sintering process. The first stage of sintering mainly forms

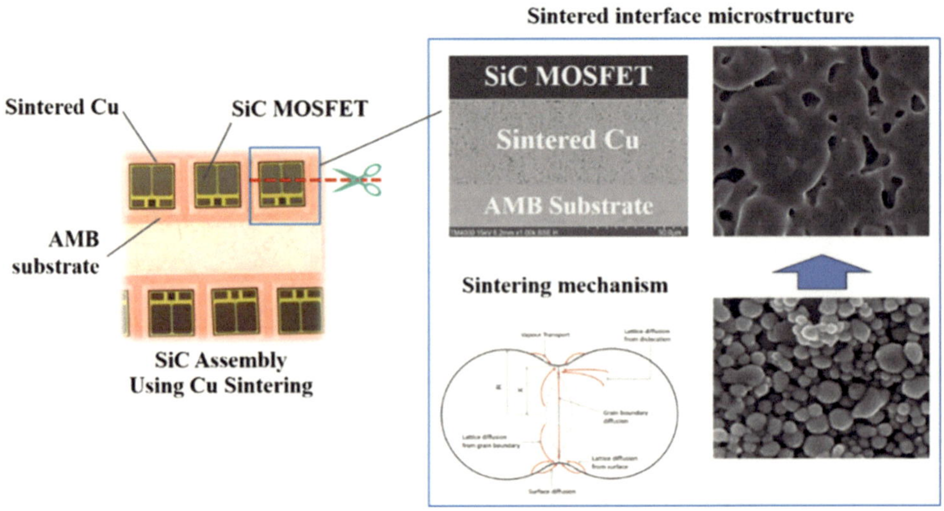

Figure 8.
Schematic diagram of various atomic diffusion paths under the equal diameter particle model.

the sintering neck through surface diffusion [39]. The subsequent two stages use grain boundary diffusion and lattice diffusion as the main diffusion mechanisms. Lattice diffusion requires higher activation energy, which means that this mechanism needs to be produced at higher sintering temperatures. Zuo et al. [40] found that during the sintering process of Cu particles below 310°C, grain boundary diffusion and surface diffusion are the main atomic diffusion paths. Lattice diffusion requires higher sintering temperatures. In the study, it was found that grain boundaries move during the sintering process, which helps reduce the porosity at the initial bonding interface and enhances the strength of the interconnection structure. In addition, they also discovered two twin formation mechanisms [41]: First, during the grain merging process, twins may be formed due to incomplete orientation; second, twin grains can also absorb nearby grains and remove them. The crystal structure transforms into twins. Liu et al. [25] confirmed the existence of high-density coherent twins inside sintered Cu through transmission electron microscopy (TEM) observations, and further found that the twin structure can significantly improve the strength and electrical properties of the material.

3.2 Low-temperature sintering effect factors

The performance of the joint after sintering is affected by many factors, such as sintering temperature, pressure, and time, which will have a significant impact on the sintering process. Yang et al. [42] studied the main factors affecting nano-silver joints, and the researchers studied many factors. The results show that when other factors are determined, as the sintering temperature or sintering pressure or sintering time increases, the shear strength of the sintered joint also increases. Under the conditions of sintering temperature and time of 300°C and 60 min, respectively, as the sintering pressure increases, the shear strength increases; under the conditions of temperature and pressure of 300°C and 15 min, respectively, as the sintering pressure increases, the shear strength increases. As the sintering time increases, the shear strength also increases; under the conditions of pressure and time of 7 MPa and 60 min, respectively, as the sintering temperature increases, the shear strength also increases. However, in practical industrial applications, people always pursue low-temperature, low-pressure, and short-time sintering processes. This is because low temperature and short sintering time can reduce sintering energy consumption and reduce cracks, deformation, and warpage caused by mismatch in thermal expansion coefficients. And excessive pressure can easily increase the probability of chip failure [18]. Therefore, many researchers explore more suitable sintering processes by changing various conditions. In addition, the thickness of the slurry layer will also affect the strength of the sintered joint. If the thickness is too large, the organic matter in the slurry will not be easily drained out during sintering, which will affect the strength. If the thickness is too thin, it will easily lead to uneven paste layers and affect the shear strength [43].

In addition, the preheating temperature can also affect the quality of the sintered joint. If the temperature is too low, there will be too much organic residue in the slurry, which will affect the connection. If the temperature is too high, the activity of the slurry will decrease and the voids will increase. Zhang et al. [44] found that the most suitable preheating temperature is 80°C when conducting joint experiments with silver micro-nano slurry. Copper paste also has good properties like silver paste, and its cost is lower than that of silver. Moreover, like silver paste, the shear strength of sintered joints of copper paste increases with the increase of sintering temperature, sintering pressure, and sintering time, and high-quality joints can be formed under low-temperature conditions.

However, copper is more easily oxidized than silver. The surface of the copper slurry will be oxidized into copper oxide, which affects the connection strength of the sintered joint, so the sintering process requires higher requirements. In order to inhibit the oxidation of copper particles, researchers usually sinter in a nitrogen, hydrogen, or formic acid atmosphere, or modify the surface by adding a coating layer and surface modification to reduce the occurrence of oxidation [45]. In addition, Zuo et al. [46] studied a rapid and pressureless in situ reduction sintering process, which can effectively reduce the oxides on the surface of copper nanoparticles and effectively inhibit the reappearance of oxides, High-strength joints with a shear strength exceeding 30 MPa can be produced at 220°C in air.

4. Advanced packaging of SiC WBG semiconductors

The packaging of SiC power devices mainly follows the following technical routes for different application scenario requirements: (1) Based on traditional silicon-based insulated gate bipolar transistor (IGBT) packaging technology, utilizing silver sintering or high-temperature brazing alloy core interconnection technology + coarse aluminum or copper wire bonding technology + single-sided thermally conductive silicone grease cooling technology + silicone gel or potting compound insulation technology, examples include HPD and XM3 [47]. HPD and XM3 refer to high-performance drive modules and communication modules. The former is commonly found in industrial automation, motor drives, or high-performance power systems, and refers to high-performance drive modules used to control motors, servo systems, or other high-precision motion control equipment. It has the characteristics of high power density and high efficiency. Communication modules, on the other hand, are used to enable data transmission and signal exchange between devices, systems, or networks. (2) Utilizing double-sided silver sintering chip interconnection technology + copper wire bonding or copper CliP technology + transfer molding encapsulation technology + single-sided direct cooling technology for the substrate/heat sink, examples include Danfoss DCM1000X and ABB RoadPAK half-bridge SiC power modules. (3) Based on silver sintering chip interconnection technology + flexible substrate low-inductance interconnection technology + single-sided direct cooling technology for the substrate/heat sink, an example is the eMPac SiC power module developed by Semikron [31]. (4) Silver sintering chip interconnection technology + double-sided direct cooling technology for the substrate/heat sink, examples include Hitachi's Card-like, Delphi's Viper, and Onsemi's VE-Trac SiC power modules with low thermal resistance [48]. (5) Based on rigid or flexible press-pack technology for traditional silicon-based high-voltage, high-capacity devices, by introducing low-temperature silver sintering technology, large-scale parallel packaging of high-voltage IGBT chips such as 4.5 and 6.5 kV is achieved, aiming to significantly reduce the cooling challenges posed by the high contact thermal resistance of traditional press-pack designs [49].

The typical application of low-temperature silver paste in power electronic modules is shown in **Figure 9**. It can be seen that the advantages of high thermal conductivity, high-temperature resistance, and support from new packaging architectures of low-temperature silver paste make it play a very important role in power packaging for voltage levels ranging from 650 V to 6.5 kV [50]. Limited by current SiC production capacity and cost factors, low-temperature silver paste is mainly used in SiC power module packaging for the main inverters of high-end new energy vehicle models [51]. However, with a significant increase in SiC production capacity, its production cost is

Figure 9.
Module packaging architecture transition from Si-based IGBT to SiC MOSFETs.

expected to decrease substantially, a trend that will greatly promote the widespread application of SiC power devices in power electronic systems. The specific applications of low-temperature sintering technology are mentioned as follows:

4.1 DTS

DTS Silicon Carbide Packaging Low-Temperature Sintering Technology is an advanced semiconductor packaging process that combines the high-performance advantages of silicon carbide (SiC) materials with the high reliability of low-temperature sintering technology. Silicon carbide, as a wide-bandgap semiconductor material, exhibits excellent properties such as high thermal conductivity, high breakdown electric field, and high electron saturation drift velocity, making it widely used in high-power, high-temperature, and high-frequency electronic devices [36]. However, traditional packaging processes often require high-temperature treatments, which may cause thermal damage to SiC devices and packaging materials. Low-temperature sintering technology effectively addresses this issue by performing sintering at relatively low temperatures (typically below 300°C) [52]. This technology utilizes low-temperature sintering materials, such as nano-silver paste, to firmly connect silicon carbide chips to substrates (e.g., copper or ceramic substrates), while providing excellent thermal conductivity and mechanical strength. Its advantages include avoiding the impact of high temperatures on device performance, improving the thermal management capabilities of packaging, enhancing the reliability and durability of connections, and meeting environmental requirements. Additionally, low-temperature sintering technology is suitable for complex packaging designs, enabling it to meet the high-performance and high-reliability demands of high-power modules (such as those used in electric vehicles and photovoltaic inverters).

Figure 10 is a schematic diagram of the Die Top System structure, from top to bottom, there are copper wires, copper foils, silver pastes, chips, silver pastes, and substrates. **Figure 2** is a comparison of the most commonly used chip top connection processes and the actual application of the Die Top System. Compared with the past, the Die Top System uses copper wires, and the connection layer has also changed from solder to silver paste. This has made significant progress in reliability and

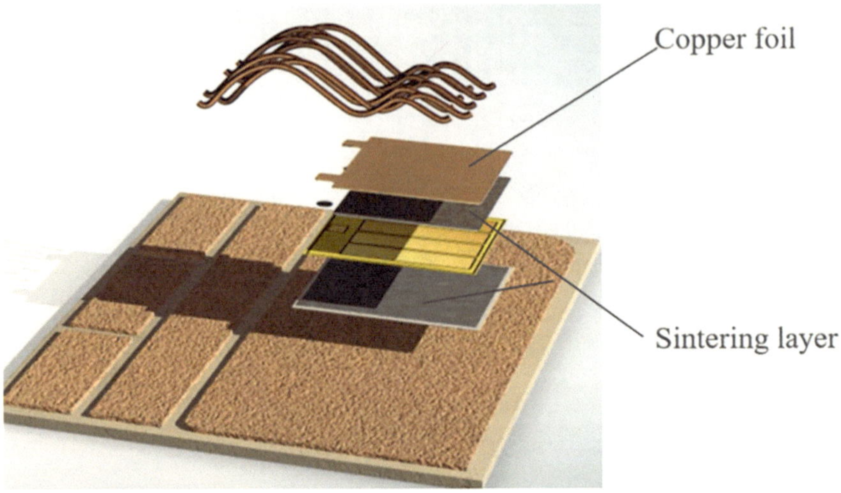

Figure 10.
Die top system structures.

current-carrying capacity. First, compared with solder chip mounting and aluminum wire bonding, highly conductive copper foil is applied to the semiconductor chip using a sintering paste. This can evenly distribute the current on the semiconductor chip, thereby reducing heat and better balancing hot spots in the semiconductor. The foil also strengthens the chip during the production of power modules. As a result, more efficient copper bonding wires can be used, instead of conventional aluminum bonding wires [53]. This increases the chip current-carrying capacity by more than 50% and extends the service life by more than 50 times. Second, Die Top System not only increases the current-carrying capacity of the power module, but also increases its power cycling capability by 10 times or more, which makes junction temperatures of up to 200°C possible. It also significantly reduces the cost per ampere, optimizing performance and profitability. Various metallized surfaces such as copper and copper alloys in the Die Top System process, with functional surfaces such as silver, gold, or palladium, have a thickness of 30–500 μm, a thermal conductivity of 180–390 W/mK, and a tensile strength of 200–650 N/mm2. The recommended pressure range for silver sintering is 10–30 MPa and the temperature range is 230–280°C.

4.2 Cu clip sintering

Compared to DTS technology, the combination of Cu Clip technology and low-temperature sintering offers significant advantages. Cu Clip utilizes highly conductive copper material, enhancing current-carrying capacity, while low-temperature sintering optimizes the conductivity of the connection interface, reducing package conduction resistance and power losses [53]. By replacing traditional bonding wires, Cu Clip simplifies the connection path, significantly lowering parasitic inductance and improving high-frequency performance and switching speed. In terms of packaging process, Cu Clip directly connects the chip to the substrate through low-temperature sintering, streamlining the process, whereas DTS requires additional copper foil or plating layers, making it more complex. The high thermal conductivity and double-sided cooling design of Cu Clip, combined with the high thermal conductivity connection layer formed by low-temperature sintering, significantly enhances heat dissipation, making it suitable for high-power-density devices [54]. Additionally, the high-strength connection layer formed by low-temperature sintering improves mechanical ad thermal cycling performance, reducing the risk of failure due to thermal expansion or vibration, thereby enhancing long-term reliability. In summary, the combination of Cu Clip technology and low-temperature sintering excels in current-carrying capacity, conduction resistance, parasitic inductance, thermal conductivity, and reliability. It simplifies the process, increases power density, efficiency, and heat dissipation, making it the preferred technology for high-power semiconductor packaging [55].

Device Level using Cu Clip: **Figure 11** shows the typical applications of Cu clip device Packaging. By replacing the thermal interface material (TIM) or solder with a thinner, highly conductive pressure sinter layer, the thermal resistance (Rth) can be reduced by 10–15%. This improvement also enhances reliability by effectively managing the stress and warpage between the package and the heat sink, as the pressure sintering process can be achieved at a lower temperature compared to other methods [56]. This advancement enables STMicroelectronics' STPAK package for automotive traction inverters to be directly mounted onto the inverter heat sinks using pressure sintering with Argomax silver sinter paste. The paste features specifically designed silver particles that work optimally with low-pressure sintering and the STPAK package materials. The Argomax silver sinter paste creates a strong bond through silver metallic bonds with direct bonded copper (DBC) substrates, providing high levels of electrical conductivity and good thermal properties. This not only improves long-term reliability but also reduces both direct conductive losses and inductive losses by placing the STPAK transistors closer together and decreasing the number of discrete devices required. As a result, a 93% efficiency in inverter designs has been achieved, which translates to an additional 10% increase in vehicle range [57–61].

Module Level using Cu clip: As depicted in **Figure 12**, in epoxy molded cases, Direct-Lead-Bonding (DLB) assemblies involve chips being soldered on one side to a common lead frame, while a dedicated copper clip is directly soldered or sintered onto the other side of the chips [62]. This configuration provides more reliable electrical connections and enhanced power cycling capability.

Figure 13 shows the newly developed next-generation SiC-based power module, the "TRCDRIVE pack™," which features a half-bridge circuit. The SiC MOSFET (metal-oxide semiconductor field-effect transistor) dies are bonded to the substrate using a silver-sintered layer, resulting in low thermal resistance [63]. To achieve lower

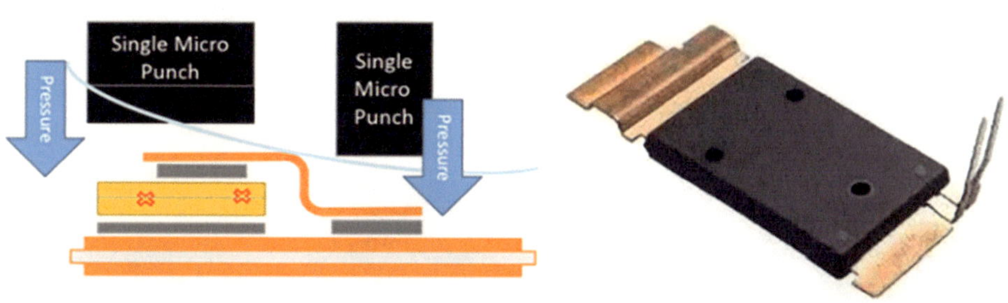

Schematic Diagram of Copper Clip Sintering

Cu clip packaging case

Figure 11.
Cu clip device packaging.

Figure 12.
Mitsubishi CT600DJH060 DLB.

Figure 13.
TRCDRIVE pack™.

total module resistance and inductance, higher power density, and enhanced reliability, copper clips are employed, instead of multiple wires or ribbons.

The BMW iX3 marks the brand's first foray into the all-electric sports utility vehicle (SUV) segment. It incorporates the fifth generation of BMW's electric powertrain technology. The power module, a 750 V, 1200 A, three-phase device manufactured by Fuji Electric Co., Ltd., is composed of Si reverse conducting IGBT (RC-IGBT) switches. RC-IGBTs integrate an IGBT and a free-wheeling diode on a single chip. **Figure 14** (left side) presents the top view of the power module [64]. **Figure 14** (middle) shows the top view of the inner configuration after the mold resin and surrounding case have been removed, with its location indicated by a red-dashed box in **Figure 14** (left side). The two RC-IGBTs on top are connected in parallel for the low side, while the two below are connected in parallel for the high side. The nodes NL-G and NL-E represent the islands where the gate and emitter of the low-side bare dies are wire bonded, respectively. Copper clips are used to connect the source pads of the

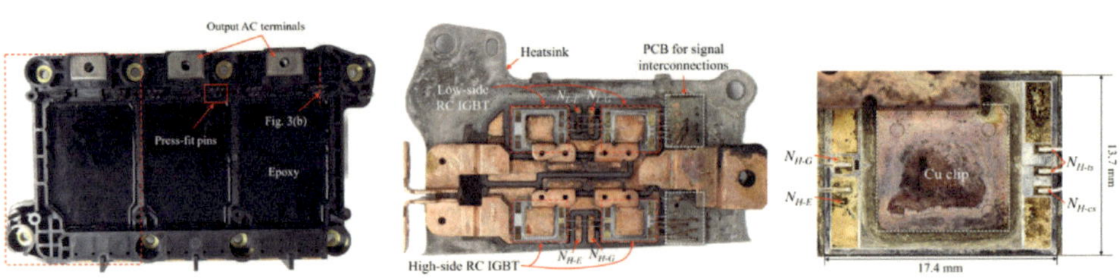

Figure 14.
Power module of BMW iX3.

bare dies to the respective terminals, as shown in **Figure 14** (right side), replacing wire bonds [65]. Wire bonds are prone to lift-off and cracking at the heel, which can cause reliability issues over the product's lifecycle. One drawback of copper clips is their thermal expansion. Encapsulating the module in mold resin mitigates this effect [66, 67].

4.3 PFC sintering

In Semikron's FPC sintering technology, the silver sintering process plays a pivotal role in connecting the flexible circuit board (flex board) to the power semiconductor chips, enabling a solderless, wire-bond-free design that enhances performance and reliability. The schematic cross-section of the SKiN device is shown in **Figure 15** [68]. This advanced sintering method involves the following steps: First, a silver paste—composed of nano- to micrometer-sized silver particles or flakes—is precisely applied to designated areas on the power side of the flex board using stencil or screen printing. These areas correspond to the top surfaces of the power chips (e.g., SiC MOSFETs, IGBTs, diodes) and critical electrical terminals [69]. The paste is also deposited on collector contact regions to ensure robust electrical and thermal pathways. Next, the flex board is meticulously aligned with the pre-sintered DBC substrate, which already hosts power chips bonded via the same low-temperature sintering process. The half-bridge module using flexible circuit board is shown in **Figure 16**. Alignment accuracy is critical to ensure proper contact between the silver paste and the chip surfaces. The assembly is then placed into a sintering press, where it undergoes a controlled thermal and mechanical treatment at approximately 250°C and 40 MPa pressure for about 1 minute [70–72]. Under these conditions, the silver particles coalesce into a dense, continuous metallic layer, forming low-resistance interconnections between the flex board and the chips.

Based on the PFC sintering technology, Semikron and Danfoss successfully developed eMPack® SiC power module, as shown in **Figure 16**. It is available in both 2-level and 3-level topologies, making it ideal for mid-power passenger vehicles as well as demanding commercial vehicle applications, with a power range spanning from 100 kW to 750 kW. Building on this concept, the eMPack MIDI version replicates these advantages in a more compact design, specifically optimized for mid- to low-power applications. The eMPack platform supports both 400 V and 800 V battery systems [73]. By integrating silicon carbide technology with fully sintered, low

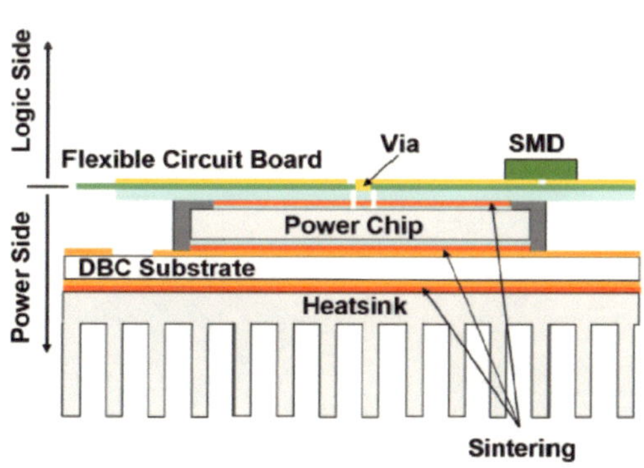

Figure 15.
Schematic cross-section of the SKiN device. The device exhibits no solder layers, nor any wire bond interconnects and is integrated with a pin fin heat sink.

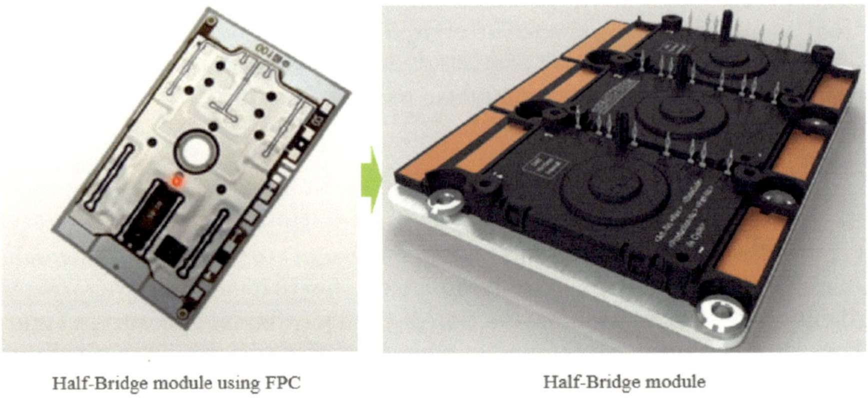

Figure 16.
eMPack® SiC power module using FPC sintering from Semikron and Danfoss.

stray inductance Direct Pressed Die Technology, it delivers exceptional power density and high reliability, setting a new standard for automotive power solutions.

4.4 Double-sided cooling packaging

The structure of the silver-sintered double-sided cooled silicon carbide module consists of SiC chips, silver-sintered connection layers, double-sided cooling substrates, and other components, as shown in **Figure 17**. Double-sided cooling (DSC) technology significantly enhances the thermal performance of the module through heat dissipation on both sides of the substrates, supporting higher power density and efficiency, while also enabling the integration of features such as temperature-sensing diodes and current-sensing IGBTs [74]. Compared to single-sided cooling modules, which suffer from high parasitic inductance (15–20 nH), high thermal resistance (0.1–0.8 K/W), and bond wire failure issues, DSC modules offer advantages, such as low parasitic inductance (less than 10 nH), up to 50% reduction in thermal resistance, a wire-bondless design, significantly extended lifespan (by an order of magnitude), and higher current-carrying capacity and power density [75]. By incorporating advanced low-temperature silver sintering technology, the thermal conductivity

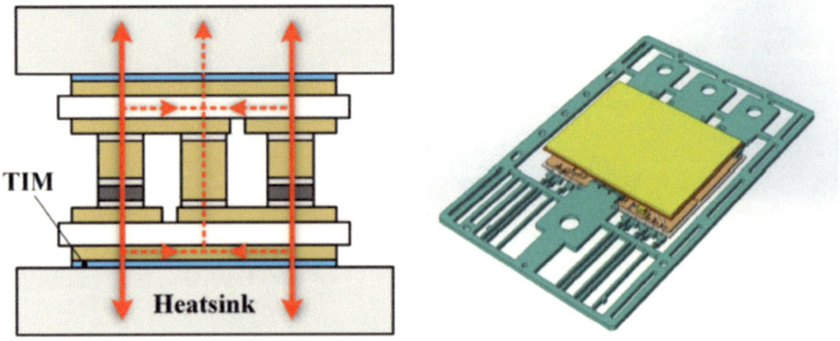

Figure 17.
The structure of DSC SiC Packaging and half-way case with Cu lead frame.

and reliability of the connection interfaces are further optimized, comprehensively improving the performance and long-term stability of the module.

The double-sided packaging silver sintering and copper sintering technologies have been widely used in the field of power module packaging, especially in new energy vehicles and high-power RF devices, demonstrating significant performance advantages and reliability improvements (**Figure 18**) [76].

The double-sided heat dissipation module of the IcyPack series of Innosilicon adopts double-sided heat dissipation, reducing the thermal resistance by about 40% and the thermal resistance as low as 0.132 K/W, which improves the outflow efficiency of the device [77, 78]. With the continuous advancement of power module packaging technology, double-sided heat dissipation technology shows great potential in applications with high-power density and high reliability [79]. By optimizing the material and structural design, the double-sided package silver sintering and copper sintering technologies further improve the heat dissipation performance and efficiency of the module, providing a more reliable solution for new energy vehicles and high-power RF devices. In addition, with the continuous development of power module packaging technology, there will be more emphasis on collaborative innovation between wide-bandgap semiconductors and packaging, as well as heterogeneous integration and functional fusion in the future [80]. For example, developing wafer level packaging for GaN and copper pillar interconnect technology for SiC, suitable for high-voltage scenarios above 1200 V [81, 82]. The development of these technologies will further promote the application of double-sided packaging silver sintering and copper sintering technologies in more fields, providing more efficient and reliable solutions for power module packaging.

4.5 Embedding packaging

In the packaging of SiC (silicon carbide) and GaN (gallium nitride) power semiconductors, Ag sintering and Cu sintering technologies are favored for their excellent performance and reliability. By building a highly conductive and thermally conductive connection layer between the bottom of the die and the substrate, these two sintering technologies not only ensure a strong bond between the die and the substrate, but also significantly improve the thermal management and electrical

Figure 18.
Innosilicon IcyPack series DSC module.

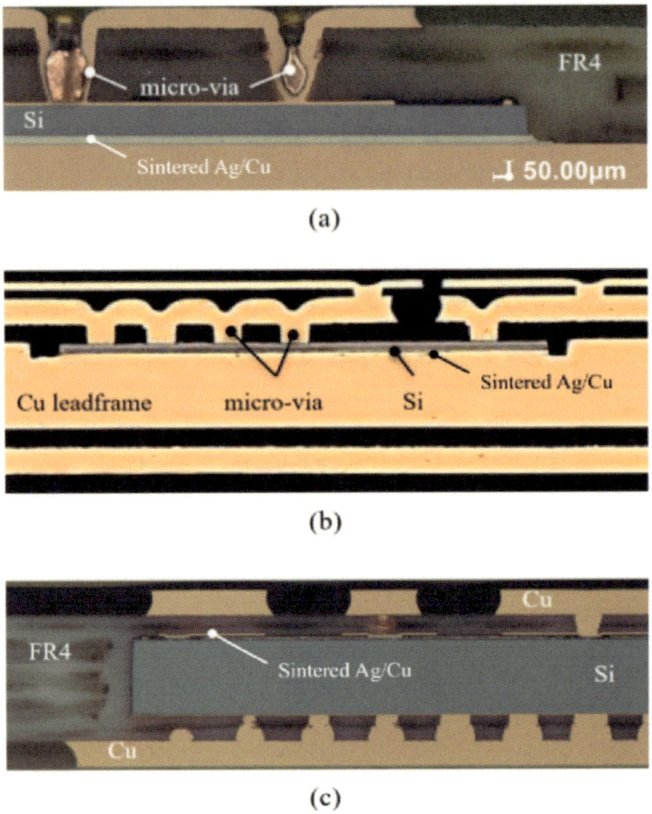

Figure 19.
Application of sintering technology in different embedding processes (a) Chip-on-substrate, (b) Chip-in-cavity [Courtesy of Schweizer electronic AG], and (c) Double-side Microvia.

performance of the entire device [83]. **Figure 19** shows the application of sintering technology in different embedding processes, including (a) Chip-on-Substrate, (b) Chip-in-Cavity, and (c) Double-side Microvia in the "Chip-on-Substrate" and "Chip-in-Cavity" embedding methods, sintered silver and sintered copper are used to secure the semiconductor die to the copper substrate, ensuring good thermal and electrical contact between the die and the substrate. In the "Double-side Microvia" embedding method, sintered silver and sintered copper are used for the backside connection of the device to improve thermal conductivity and mechanical stability [84].

Sintered silver technology plays an important role in die-attach and partial interconnect processes. It is employed at high temperatures of 230–260°C to induce a tight bond between the silver particles [85]. This high-temperature sintering process not only ensures the high strength and conductivity of the connecting layers, but also significantly reduces the thermal resistance. Fraunhofer IZM (Fraunhofer Institute for Reliability and Microintegration, Germany) has developed a power module based on Al_2O_3 DBC (aluminum oxide directly bonded to copper), which employs the sintered silver technology for chip mounting [86]. The high electrical and thermal conductivity of sintered silver enables it to effectively reduce thermal resistance, thereby significantly improving the thermal performance and overall reliability of the module. ABB has also developed a power module based on Cu lead frames using sintered silver technology. After 1 million temperature cycling tests, the module did not experience any failures, fully demonstrating the excellent reliability of sintered silver technology under extreme temperature changes.

Sintered copper technology is also widely used in die placement and partial interconnect processes. Similar to silver sintering, copper sintering needs to be carried out at temperatures between 230°C and 260°C to ensure a stable bond between the copper particles. The University of Applied Sciences Kempten has developed a power module based on Cu-IMS (copper-insulated-metal-substrate), which utilizes sintered copper technology for chip placement [87]. Sintered copper not only provides excellent electrical and thermal conductivity, but also effectively reduces stresses due to mismatched coefficients of thermal expansion. Fraunhofer IZM has also developed a power module based on Cu lead frames using sintered copper for die placement. In high- and low-temperature cycling tests, the sintered copper demonstrated good mechanical stability and electrical properties, significantly reducing the stress caused by differences in the coefficients of thermal expansion, and further enhancing the reliability and performance of the device. In summary, the application of sintered silver and sintered copper technologies in SiC and GaN power semiconductor packages not only improves the thermal and electrical properties of the devices, but also significantly enhances their reliability under extreme temperature conditions. The wide application of these technologies provides solid technical support for the development of high-performance power electronic devices.

4.6 Press-Pak packaging

Press-Pak packaging is a press structure that combines rigidity and elasticity [88], as shown in **Figure 20**. **Figure 20(a)** is a vertical structure diagram of the Press-Pak IGBT device, and **Figure 20(b)** is a distribution diagram of the chip parts of one of the structures. It consists of springs, chips, silver and metal frames. Its principle is to achieve electrical and mechanical connections through mechanical pressure without welding or other auxiliary materials [89]. Its core lies in the use of the elastic

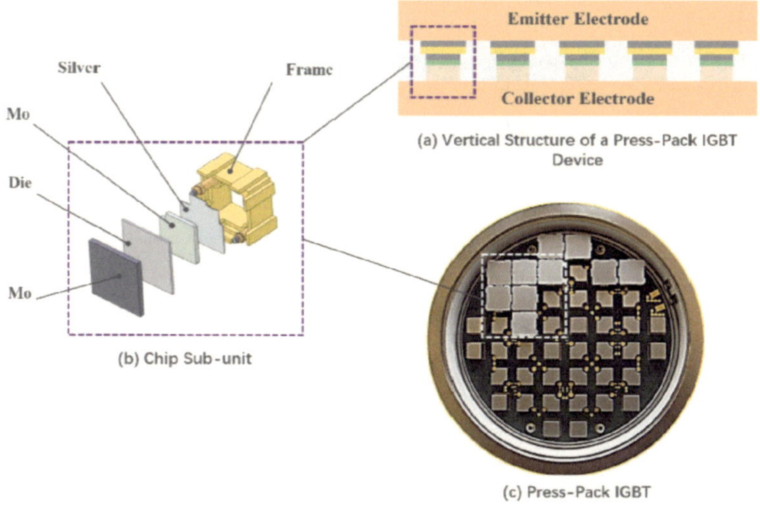

Figure 20.
Press-Pak packaging.

deformation of the pins and the close contact between the holes of the substrate to form a reliable connection. **Figure 20(c)** is a physical diagram of the IGBT chip used by Press-Pak. The core features of Press-Pak packaging are no thermal stress, high reliability, elastic adaptability, and detachability, so it is widely used in high-frequency, high-power, and high-reliability fields, such as power modules, semiconductor devices, LED packaging, and other occasions [90].

4.7 Large-area sintering

With the rapid development of high-power electronic devices, the research topic of large-area bonding in the field of electronics has received more and more attention. Large-area chip bonding (>10 mm × 10 mm), high heat dissipation power base/heat sink bonding (connecting devices to heat sinks/substrates), etc. [91], have triggered the demand for reliable large-area bonding technology with excellent heat dissipation performance. Previously, traditional thermal conductive interface materials such as thermal grease and thermal conductive adhesive were used between the substrate and the heat sink, but after using these materials, there is still thermal resistance at the interface, which affects the heat dissipation effect [92]. In addition, its long-term stability is poor, and there are aging problems and pump-out effects. For example, some TIMs (such as thermal grease) will dry up or fail over time, and under thermal cycles, TIMs may be squeezed out of the interface, reducing performance. The high conductivity of silver can reduce interface resistance and improve electrical performance, and silver sintering forms a strong metal bond, which can enhance mechanical strength, and the silver-sintered layer is not easy to age and has long-term stable performance [93]. Therefore, large-area silver sintering is becoming more and more popular. Copper and silver have similar properties, and copper has better anti-electromigration performance than silver, and the cost is lower, so large-area copper sintering is being studied more and more widely [94].

As shown in **Figure 21**, it is a 1200 V-600A SiC power module using large-area copper sintering from Zhejiang University (ZJU) [95]. **Figure 21** (left) is its vertical structure, using SiC MOSFET chip, the top of the chip is connected with DTS copper wire, and copper sintering is used at the top and bottom of the chip, and between the substrate and the heat sink [96]. Large-area sintering will occupy an important position in the future large-area connection due to its high thermal conductivity, high electrical conductivity, high-temperature stability, high mechanical strength, low thermal resistance, environmental protection, and other characteristics. Among them, copper will have a bright application prospect due to its cost advantage.

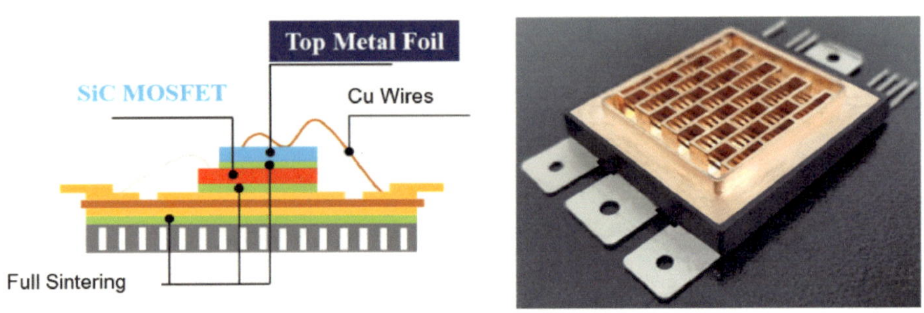

Figure 21.
1200 V-600A SiC module using large-area Cu sintering from Zhejiang University (ZJU).

5. Pressure auxiliary sintering system

Sintering does not require many equipment, Sinterstar innovate-F-XL can realize many sintering applications, such as chip substrate sintering (copper area and non-copper area), clip sintering, lead frame sintering, 1-up package sintering on cooling plate, double-sided cooling unit (all sintering steps), sintering substrate on cooling plate, and Mo-chip-Mo sintering. As shown in **Figure 22**, different sintering in the left frame diagram can be completed by Sinterstar innovate-F-XL. And the O2 level in Sinterstar innovate-F-XL can reach <50 ppm, which is very suitable for sintering copper area products, such as bare copper lead frame, copper substrate (DBC and active metal brazing (AMB)). Its principle is to evacuate the single polytetrafluoroethylene (PTFE) film on the loading fixture in a microcontrolled environment, and form a controlled O_2 environment in the shortest time by blowing N_2. Its controlled environment is active during the whole cycle, including preheating, sintering, and cooling [96].

Cu clip packaging has the advantages of high-quality connection, excellent electrical and thermal conductivity, high packaging reliability, high material utilization, and fast production efficiency, and is increasingly valued [97]. **Figure 23** shows the sintering process of Cu clip, which starts with the attachment of the sintering film, followed by the attachment and sintering of the chip, followed by the dispersion of the sintering slurry, and finally the sintering of the copper clip. The chip sintering can be completed by Sinterstar innovate-F-XL, and the microjoint can be completed by AMX instruments. The above is a brief introduction to the equipment commonly used in the sintering process.

6. Summary

This paper offers a brief yet comprehensive overview of the sintering mechanisms, fabrication techniques, and cutting-edge applications of low-temperature silver/copper sintering materials in modern wide-bandgap semiconductor packaging. The content serves as a reference for postgraduate students, engineers, and scholars.

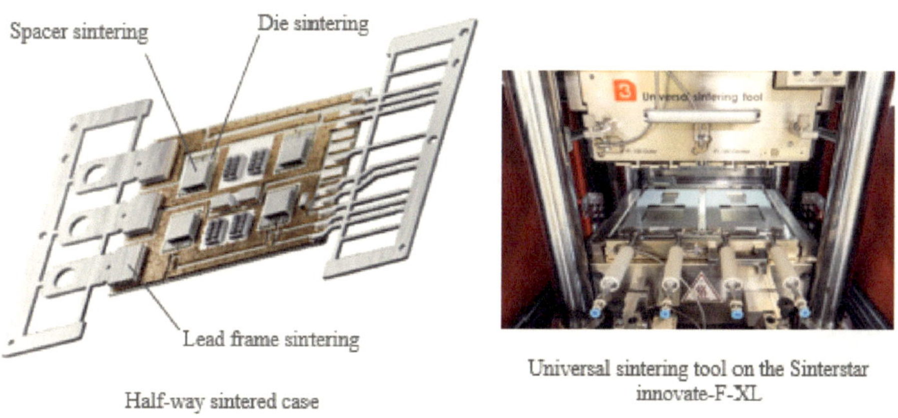

Figure 22.
Full Sintering packaging.

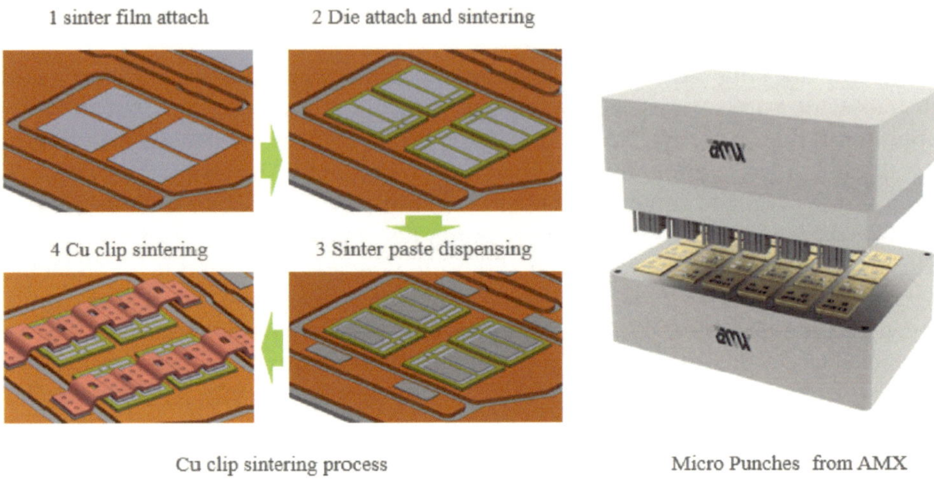

Figure 23.
Full Sintering packaging case.

Acknowledgements

This work acknowledges Dr. Rui Ding from Dongfang Electric Academy of Science and Technology Co., Ltd. and Mr. Shuai Shi, a chief engineer at the National New Energy Vehicle Technology Innovation Center (NEVC), for their valuable input. Funding support was provided by the National Key Research and Development Program of China under Grant 2023YFB3609504.

Author details

Haidong Yan[1]*, Wanli Li[2], Yaqin Liao[3] and Chaohui Liu[4]

1 College of Electrical Engineering, Zhejiang University, Hangzhou, China

2 School of Mechanical Engineering, Jiangnan University, Wuxi, China

3 Dongfang Electric Academy of Science and Technology Co. Ltd, Chengdu, China

4 National New Energy Vehicle Technology Innovation Center (NEVC), Beijing, China

*Address all correspondence to: haidong_yan@zju.edu.cn

References

[1] Alzoubi K, Lu S, Sammakia B, Poliks M. Experimental and analytical studies on the high cycle fatigue of thin film metal on PET substrate for flexible electronics applications. IEEE Transactions on Components, Packaging and Manufacturing Technology. 2011;**1**(1):43-51. DOI: 10.1109/TCPMT.2010.2100911

[2] Liu C-K et al. Development of packaging technologies for SiC power module. In: International Conference on Microsystems, Packaging, Assembly and Circuits Technology (IMPACT). 2015

[3] Lee H, Smet V, Tummala R. A review of SiC power module packaging technologies: Challenges, advances, and emerging issues. IEEE Journal of Emerging and Selected Topics in Power Electronics. 2020;**8**(1):239-255. DOI: 10.1109/JESTPE.2019.2951801

[4] Bajwa AA, Qin Y, Reiner R, et al. Assembly and packaging technologies for high-temperature and high-power GaN devices. IEEE Transactions on Components, Packaging and Manufacturing Technology. 2015;**5**(10):1402-1416

[5] Chen C, Luo F, Kang Y. A review of SiC power module packaging: Layout, material system and integration. CPSS Transactions on Power Electronics and Applications. 2017;**2**(3):170-186. DOI: 10.24295/CPSSTPEA.2017.00017

[6] Ren H et al. High-reliability wireless packaging for high-temperature SiC power device sintered by novel organic-free nanomaterial. IEEE Transactions on Components, Packaging and Manufacturing Technology. 2020;**10**(12):1953-1959. DOI: 10.1109/TCPMT.2020.3038430

[7] Braunwarth L, Amrhein S, Schreck T, et al. Ecological comparison of soldering and sintering as die-attach technologies in power electronics. Journal of Cleaner Production. 2015;**102**:408-417

[8] Yoshikawa H. Cu Paste Sinter Joining for Die-Attach of High TJ Power Devices Realized by PEG Solvent Optimization. Europe: PCIM; 2016

[9] Chen Z, Gengenbach U, Koker L, et al. Systematic investigation of novel, controlled low-temperature sintering processes for inkjet printed silver nanoparticle ink. Small. 2024;**20**(21):2306865

[10] Chen C, Sekiguchi T, Suganuma K. Bare Cu bonding by Cu-Ag composite paste in low temperature low pressure air sintering. In: 2024 International Conference on Electronics Packaging (ICEP). New Jersey, USA: IEEE; 2024. pp. 79-80

[11] Wang T, Chen X, Lu GQ, et al. Low-temperature sintering with nano-silver paste in die-attached interconnection. Journal of Electronic Materials. 2007;**36**:1333-1340

[12] Chen Z et al. Pressureless silver sintering for SiC power module interconnects: Kinetics and microstructure evolution. Journal of Materials Science. 2024;**59**(3):1125-1138

[13] Wang Y et al. Multi-phase sintering of Ag-Cu composite pastes for high-temperature power electronics. IEEE Transactions on Components, Packaging and Manufacturing Technology. 2023;**13**(2):345-356

[14] Liu J, Suganuma K. Ultrafast low-temperature sintering of Cu

nanoparticles for power electronics: Time-temperature tradeoffs. Materials & Design. 2022;**215**:110502

[15] Wang Y, Yang D, Wu P, et al. Study on low temperature sintering mechanism and performance of multiscale silver paste. In: 2023 24th International Conference on Electronic Packaging Technology (ICEPT). New Jersey, USA: IEEE; 2023. pp. 1-4

[16] Nakako H. Superior Bonding Reliability of Sintered Cu Bonding at Power Cycle Test. Europe: PCIM; 2020

[17] Zuo Y, Carter-Searjeant S, Green M, Mills L, Mannan SH. High bond strength Cu joints fabricated by rapid and pressureless in situ reduction-sintering of Cu nanoparticles. Materials Letters. 2020;**276**:128068

[18] Zhao S, Dai Y, Qin F, Li Y, An T, Gong Y. Effect of surface finish metallization layer on shearing fracture toughness of sintered silver bonded joints. Engineering Fracture Mechanics. 2022;**264**:108431

[19] Wu Z, Liu W, Feng J, et al. Novel Cu@ Ag micro/nanoparticle hybrid paste and its rapid sintering technique via electromagnetic induction for high-power electronics. ACS Omega. 2023;**8**(34):31021-31029

[20] Mehr MY, Bahrami A, Fischer H, et al. An overview of scanning acoustic microscope, a reliable method for non-destructive failure analysis of microelectronic components. In: 16th International Conference on Thermal, Mechanical and Multi-Physics Simulation and Experiments in Microelectronics and Microsystems. New Jersey, USA: IEEE; 2015. pp. 1-4

[21] Song R, Yuan F, Su Y, et al. Infrared radiation-assisted non-pressure sintering of Micron-sized silver for power electronic packaging. Electronics. 2024;**13**(8):1492

[22] Nakako H. Sintering Copper Die-Bonding Paste Curable under Pressureless Conditions. Europe: PCIM; 2017

[23] Rabay B. New Die Attach Materials: Silver and Silver/Copper Sintering Pastes. Europe: PCIM; 2022

[24] Pan X. Study on Preparation and Application of Nano-Copper Powder for Power Semiconductor Device Packaging. Europe: PCIM; 2019

[25] Liu J, Chen H, Ji H, Li M. Highly conductive Cu–Cu joint formation by low-temperature sintering of formic acid-treated Cu nanoparticles. ACS Applied Materials & Interfaces (Washington, DC: American Chemical Society). 2016;**8**(48):33289-33298

[26] Park B, Saito M, Mizuno J, Nishikawa H. Robust shear strength of Cu–Au joint on Au surface-finished Cu disks by solid-state nanoporous Cu bonding. Microelectronic Engineering. 2022;**260**:111698

[27] Yamauchi S. The Characteristics of Sintered Copper Layer by Pressure Sintering Process for Power Module Application. Europe: PCIM; 2019

[28] Nakako H. Copper Sintering Pastes for Die Bonding. Europe: PCIM; 2021

[29] Nakako H. Bonding between a Ceramic Wiring Board and Cooling Plate Using Copper Sintering Paste. Europe: PCIM; 2022

[30] Dai J, Wang Y, Grant T, Morshed M. Low-temperature die attachment by Pressureless Cu sintering for semiconductor packaging.

Journal of Electronic Materials. 2023;**52**(11):7607-7613

[31] Zhao Y, Tolla B, Katze D, et al. A hybrid pressureless silver sintering technology for high-power density electronics. Journal of Microelectronic & Electronic Packaging. 2023;**20**(3)

[32] Dai C, Wang Y, Lin P, et al. Achieving high-quality nano-silver joints for high-reliability power electronics via pressureless sintering. Authorea Preprints. 2023

[33] Wang L, Lei Z, Liang R, et al. A new SiC power module assembly based on silver sintering bonding, 2023 25th European conference on power electronics and applications (EPE'23 ECCE Europe). IEEE. 2023:1-8

[34] Wu H, Wang J, Yu S. The packaging design of SiC power modules based on silver sintering technology. In: 2024 IEEE 7th International Electrical and Energy Conference (CIEEC). New Jersey, USA: IEEE; 2024. pp. 4815-4819

[35] Hwang B-U. Pressureless transient liquid phase sintering bonding using SAC305 with hybrid Ag particles for high–temperature packaging applications. In: IEEE 70th Electronic Components and Technology Conference (ECTC). 2020. pp. 1-7

[36] Liu X, Li S, Fan J, Jiang J, Liu Y, Ye H, et al. Microstructural evolution, fracture behavior and bonding mechanisms study of copper sintering on bare DBC substrate for SiC power electronics packaging. Journal of Materials Research and Technology. 2022;**19**:1407-1421

[37] Ding L, Davidchack RL, Pan J. A molecular dynamics study of sintering between nanoparticles. Computational Materials Science. 2009;**45**(2):247-256

[38] Wang Y, Xu D, Yan H, Li C-F, Chen C, Li W. Low-temperature copper sinter-joining technology for power electronics packaging: A review. Journal of Materials Processing Technology. 2024;**332**:117971

[39] Mavliev R, Rhoades R. Selective copper metallization for advanced packaging. In: 2020 International Wafer Level Packaging Conference (IWLPC). San Jose, CA, USA; 2020. pp. 1-7. DOI: 10.23919/ IWLPC52010.2020.9375858

[40] Zuo Y, Zhao C, Robador A, Wickham M, Mannan SH. Quasi-in-situ observation of the grain growth and grain boundary movement in sintered Cu nanoparticle interconnects. Acta Materialia. 2022;**236**:118187

[41] Wu Y, Zou G, Wang S, Guo W, Zhang H, Liu L. Interfacial bonding mechanisms of sintered Cu nanoparticles on different metallization surfaces. Surfaces and Interfaces. 2024;**46**

[42] Yang H, Zhu W. Study on the main influencing factors of shear strength of nano-silver joints. Journal of Materials Research and Technology. 2020;**9**(3):4133-4138

[43] Hou F, Sun Z, Su M, Fan J, You X, Li J, et al. Review of die-attach materials for SiC high-temperature packaging. IEEE Transactions on Power Electronics. 2024;**39**(10):13471-13486

[44] Zhang X, Zhang Y, Wang C, Zhu P, Xiang B, Zhao T, et al. Exploration of key factors for the sintering of micro-Nano silver paste. In: 2022 23rd International Conference on Electronic Packaging Technology (ICEPT). Piscataway, NJ, USA: IEEE; 2022. pp. 1-6

[45] Kobayashi Y, Shirochi T, Maeda T, Yasuda Y, Morita T. Microstructure of

metallic copper nanoparticles/metallic disc interface in metal–metal bonding using them. Surface and Interface Analysis. 2013;**45**(9):1424-1428

[46] Yoon J-W, Back J-H, Jung S-B. Effect of surface finish metallization on mechanical strength of Ag sintered joint. Microelectronic Engineering. 2018;**198**:15-21

[47] Ermantraut E et al. New process for selective additive metallization of alumina ceramic substrates. IEEE Transactions on Components, Packaging and Manufacturing Technology. 2019;**9**(1):138-145. DOI: 10.1109/TCPMT.2018.2881410

[48] Ding C, Lu S, Zhang Z, Zhang K, Nguyen T, Ngo KDT, et al. Double-side cooled SiC MOSFET power modules with sintered-silver interposers for a 100-kW/L traction inverter. IEEE Transactions on Power Electronics. 2023;**38**(8):9685-9694

[49] Mingxiang, Chen. Method for Preparing Metallized Ceramic Substrate by Low-Temperature Sintering; 2012

[50] Brinkhues S, Kanthamneni A, Brose A, Majcherek S, Schmidt B. Investigation of adhesion strength of metallization on thermoplastic and ceramic substrates. In: 2016 12th International Congress Molded Interconnect Devices (MID). Wuerzburg, Germany: IEEE; 2016. pp. 1-6. DOI: 10.1109/ICMID.2016.7738935

[51] Goth C, Kuhn T, Gion G, Franke J. Hot pin pull method - new test procedure for the adhesion measurement for 3D-MID. Advanced Materials Research. 2014;**1038**:115-120

[52] Hsiu-hung C, Yuan Z, Chung-Lung C. Experimental study of coefficient of thermal expansion of aligned graphite thermal interface materials. Frontiers in Heat and Mass Transfer. 2013;**4**. DOI: 10.5098/HMT.V4.1.3004

[53] Molenaar M, Kardan F, Shekhar A, Bauer P. Power and thermal cycling testbed for end of life assessment of semiconductor devices. In: 2023 IEEE IECON. IEEE; 2023. pp. 1-6. DOI: 10.1109/iecon51785.2023.10312731

[54] Wang L et al. Cu clip-bonding method with optimized source inductance for current balancing in multichip SiC MOSFET power module. IEEE Transactions on Power Electronics. 2022;**37**(7):7952-7964. DOI: 10.1109/TPEL.2022.3141373

[55] Lwin KK, Tubillo CE, Dimaano Panumard TJ, Suthiwongsunthorn N, Sirinorakul S. Copper clip package for high performance MOSFETs and its optimization. In: 2016 IEEE 18th Electronics Packaging Technology Conference (EPTC), Singapore. 2016. pp. 123-128. DOI: 10.1109/EPTC.2016.7861457

[56] Gilfrich JV, Birks LS. Spectral distribution of X-ray tubes for quantitative X-ray fluorescence analysis. Analytical Chemistry. 1968;**40**(7):1077-1080

[57] Aryan P, Sampath S, Sohn H. An overview of non-destructive testing methods for integrated circuit packaging inspection. Sensors. 2018;**18**(7):1981

[58] Vaga R, Bryant K. Recent advances in x-ray technology. In: 2016 Pan Pacific Microelectronics Symposium (Pan Pacific). Piscataway, NJ, USA: IEEE; 2016. pp. 1-10

[59] Rudzki J, Jensen L, Poech M, et al. Quality evaluation for silver sintering layers in power electronic modules. In: 2012 7th International Conference on

Integrated Power Electronics Systems (CIPS). Piscataway, NJ, USA: IEEE; 2012. pp. 1-6

[60] Martz HE, Logan CM, Schneberk DJ, et al. X-Ray Imaging: Fundamentals, Industrial Techniques and Applications. Boca Raton, Florida, USA: CRC Press; 2016

[61] Dwivedi SK, Vishwakarma M, Soni A. Advances and researches on non destructive testing: A review. Materials Today: Proceedings. 2018;**5**(2):3690-3698

[62] Gilmore RS, Tam KC, Young JD, et al. Acoustic microscopy from 10 to 100 MHz for industrial applications. Philosophical Transactions of the Royal Society of London. Series A, Mathematical and Physical Sciences. 1986;**320**(1554):215-235

[63] Schmid M, Bhogaraju SK, Liu E, et al. Comparison of nondestructive testing methods for solder, sinter, and adhesive interconnects in power and opto-electronics. Applied Sciences. 2020;**10**(23):8516

[64] Maldague X. Theory and practice of infrared technology for nondestructive testing[J]. Wiley-Interscience; 2001

[65] Wiecek B. Review on thermal image processing for passive and active thermography. Conference proceedings: Annual international conference of the IEEE engineering in medicine and biology society. IEEE Engineering in Medicine and Biology Society. Conference. 2005;**1**:686-689

[66] Usamentiaga R, Venegas P, Guerediaga J, et al. Infrared thermography for temperature measurement and non-destructive testing. Sensors. 2014;**14**(7):12305-12348

[67] Zeh K. Selection and qualification of a test method for determining adhesion of MID metallization in the LDS technology. In: 10th International Congress Molded Interconnect Devices Research Association Molded Interconnect Devices 3-D MID e.V Fürth. 2012

[68] Stockmeier T, Beckedahl P, Göbl C, Malzer T. SKiN: Double side sintering technology for new packages. In: 2011 IEEE 23rd International Symposium on Power Semiconductor Devices and ICs, San Diego, CA, USA. 2011. pp. 324-327. DOI: 10.1109/ISPSD.2011.5890856

[69] Ibarra-Castanedo C, Galmiche F, Darabi A, et al. Thermographic nondestructive evaluation: Overview of recent progress. Thermosense XXV. 2003;**5073**:450-459

[70] Dreher P, Schmidt R, Vetter A, et al. Non-destructive imaging of defects in Ag-sinter die attach layers–a comparative study including X-ray, scanning acoustic microscopy and thermography. Microelectronics Reliability. 2018;**88**:365-370

[71] Göbl C, Faltenbacher J. Low temperature sinter technology die attachment for power electronic applications, 2010 6th international conference on integrated power electronics systems. IEEE. 2010:1-5

[72] Dresel F, Letz S, Zischler S, et al. Selective silver sintering of semiconductor dies on PCB, PCIM Europe 2018. In: International Exhibition and Conference for Power Electronics, Intelligent Motion, Renewable Energy and Energy Management. Berlin, Germany: VDE; 2018. pp. 1-8

[73] Müller J, Letz SA, Simon FB, et al. Silver sintering of packaged GaN-devices on printed circuit board. Journal of Microelectronics and Electronic Packaging. 2022;**19**(1):18-24

[74] Liu C-K et al. Double-sided cooling SiC power module packaging for industrial motor driving system. In: 2020 15th International Microsystems, Packaging, Assembly and Circuits Technology Conference (IMPACT), Taipei, Taiwan. New Jersey, USA: IEEE; 2020. pp. 105-108. DOI: 10.1109/IMPACT50485.2020.9268577

[75] Akbar SQUA, Sekar AK, Cerezo J. Performance evaluation of CoolSiC CoolSiC (TM) 2 kV SiC MOSFET discrete in 1500 V DC link systems. In: PCIM Europe 2024; International Exhibition and Conference for Power Electronics, Intelligent Motion, Renewable Energy and Energy Management. Nürnberg, Germany: Mesago PCIM GmbH; 2024. pp. 2036-2043. DOI: 10.30420/566262286

[76] Paul R, Hassan A, Mantooth HA. A double-sided cooled power module with embedded decoupling capacitors. IEEE Journal of Emerging and Selected Topics in Power Electronics. 2024;**12**(2):1813-1821. DOI: 10.1109/JESTPE.2023.3348696

[77] Liu C-K et al. Development of packaging technologies for SiC power module. In: 2015 10th International Microsystems, Packaging, Assembly and Circuits Technology Conference (IMPACT). Taipei, Taiwan: IEEE; 2015. pp. 269-272. DOI: 10.1109/IMPACT.2015.7365232

[78] Liu C-K, Tzeng C-M, Fang R-C, Li C-W, Goo J. SiC power module for motor driving system. In: 2018 International Conference on Electronics Packaging and iMAPS all Asia Conference (ICEP-IAAC). Mie, Japan: IEEE; 2018. pp. 363-366. DOI: 10.23919/ICEP.2018.8374324

[79] Liebig S, Nuber M, Kriegel K, Weidner K. Ultra-compact SiC power module with sintered DCB on microchannel cooler. In: Proceedings of PCIM Europe 2015; International Exhibition and Conference for Power Electronics, Intelligent Motion, Renewable Energy and Energy Management. Nuremberg, Germany: Mesago PCIM GmbH; 2015. pp. 1-6

[80] Lu GQ, Mei Y, Wang M, et al. Low-temperature silver sintering for bonding 3D power modules. In: 2019 6th International Workshop on Low Temperature Bonding for 3D Integration (LTB-3D). New Jersey, USA: IEEE; 2019. pp. 19-19

[81] Emon AI, Carlton H, Harris J, et al. Design and optimization of 650V/60A double-sided cooled multichip GaN module. In: 2021 IEEE Applied Power Electronics Conference and Exposition (APEC). New Jersey, USA: IEEE; 2021. pp. 2313-2317

[82] Paul R, Alizadeh R, Li X, Chen H, Wang Y, Mantooth HA. A double-sided cooled SiC MOSFET power module for EV inverters. IEEE Transactions on Power Electronics. 2024;**39**(9):11047-11059

[83] Duan G, Kanaoka Y, McRee R, Nie B, Manepalli R. Die embedding challenges for EMIB advanced packaging technology. In: 2021 IEEE 71st Electronic Components and Technology Conference (ECTC). San Diego, CA, USA: IEEE; 2021. pp. 1-7. DOI: 10.1109/ECTC32696.2021.00012

[84] Boettcher L, Manessis D, Ostmann A, Karaszkiewicz S, Reichl H. Embedding of chips for system in package realization - technology and applications. In: 2008 3rd International Microsystems, Packaging, Assembly & Circuits Technology Conference, Taipei, Taiwan. IEEE; 2008. pp. 383-386. DOI: 10.1109/IMPACT.2008.4783892

[85] Lu S, Zhao T, Burgos R, et al. Packaging of (650 V, 150 a) GaN HEMT

with low parasitics and high thermal performance. In: 2021 International Conference on Electronics Packaging (ICEP). New Jersey, USA: IEEE; 2021. pp. 39-40

[86] Dong Z, Fan Y, Yan H, et al. HEMT integrated power module with silver sintering processes. In: 2023 IEEE 2nd International Power Electronics and Application Symposium (PEAS). New Jersey, USA: IEEE; 2023. pp. 843-848

[87] Alderman A, Burgyan LL, Narveson B, Parker E. 3-D embedded packaging technology: Analyzing its needs and challenges. IEEE Power Electronics Magazine. 2015;**2**(4):30-39. DOI: 10.1109/MPEL.2015.2485359

[88] Tong J, Huang AQ. Design and characterization of a novel double-side cooled press-pack packaging structure. In: 2023 IEEE Energy Conversion Congress and Exposition (ECCE), Nashville, TN, USA. IEEE; 2023. pp. 5480-5484. DOI: 10.1109/ECCE53617.2023.10362912

[89] Hermansson W, Chimento F, Jonsson T. Robustness evaluation of high voltage press pack IGBT modules in enhanced short circuit test. In: 2010 IEEE Energy Conversion Congress and Exposition, Atlanta, GA, USA. 2010. pp. 92-99. DOI: 10.1109/ECCE.2010.5618068

[90] Jeon J, Seong J, Lim J, et al. Finite element and experimental analysis of spacer designs for reducing the thermomechanical stress in double-sided cooling power modules. IEEE Journal of Emerging and Selected Topics in Power Electronics. 2020;**9**(4):3883-3891

[91] Chen C, Kim D, Wang Z, Zhang Z, Gao Y, Choe C, et al. Low temperature low pressure solid-state porous Ag bonding for large area and its high-reliability design in die-attached power modules. Ceramics International. 2019;**45**(7):9573-9579

[92] Chen C, Noh S, Zhang H, Choe C, Jiu J, Nagao S, et al. Bonding technology based on solid porous Ag for large area chips. Scripta Materialia. 2018;**146**:123-127

[93] Xie Y, Wang Y, Mei Y, Xie H, Zhang K, Feng S, et al. Rapid sintering of nano-Ag paste at low current to bond large area (>100 mm2) power chips for electronics packaging. Journal of Materials Processing Technology. 2018;**255**:644-649

[94] Zhang B, Chen C, Sekiguchi T, Liu Y, Li C, Suganuma K. Development of anti-oxidation Ag salt paste for large-area (35 × 35 mm2) Cu-Cu bonding with ultra-high bonding strength. Journal of Materials Science & Technology. 2022;**113**:261-270

[95] Haidong Y, Yehui M, Yuncan L, et al. Characterization of electrothermal performance of high-power silicon carbide modules based on low-temperature copper sintering technology. Electronics & Packaging. 2025;**25**(3):78-83

[96] Noh S, Zhang H, Yeom J, Chen C, Li C, Suganuma K. Large-area die-attachment by silver stress migration bonding for power device applications. Microelectronics Reliability. 2018;**88-90**:701-706

[97] Li J, Li X, Wang L, Mei Y-H, Guo-Quan L. A novel multiscale silver paste for die bonding on bare copper by low-temperature pressure-free sintering in air. Materials & Design. 2018;**140**:64-72. ISSN 0264-1275.

Chapter 5

Energy Storage with Superconducting Magnets: Low-Temperature Applications

Ankit Anand, Abhay Singh Gour, Tripti Sekhar Datta and Vutukuru Vasudeva Rao

Abstract

Superconducting Magnet Energy Storage (SMES) systems are utilized in various applications, such as instantaneous voltage drop compensation and dampening low-frequency oscillations in electrical power systems. Numerous SMES projects have been completed worldwide, with many still ongoing. This chapter will provide a comprehensive review of SMES projects around the globe, detailing the methodologies for maintaining the low temperatures required for these devices. Superconductors have zero joule loss below their critical temperature, allowing SMES to save energy without any loss. Additionally, since there is no mechanical conversion when supplying energy, SMES systems boast very high efficiency, up to 95%. Potential new applications of SMES will be thoroughly explored, and the significant issues facing current technology, as well as possible advancements in this field, will be discussed.

Keywords: SMES, low temperature, superconducting magnet, energy storage, cooling methods

1. Introduction

Energy Storage Systems (ESS) are required for constant power supply, which can store energy during surplus supply and be used when power production is deficient. **Figure 1** shows the major ESS used presently in the globe. The energy storage landscape features various technologies, each suited for specific needs. Mechanical systems, like pumped hydro, flywheel, and compressed air, store energy as motion or potential energy. Thermal systems store heat energy, while electrostatic systems, like supercapacitors, use electric fields. Chemical systems, including fuel cells and biofuels, convert chemical energy into electricity. Electrochemical systems, such as lead-acid and Li-ion batteries, rely on chemical reactions. Magnetic systems, especially Superconducting Magnet Energy Storage (SMES), store energy in magnetic fields, offering quick response and high efficiency. This makes SMES a key player in advancing energy storage solutions.

Superconducting Magnet Energy Storage (SMES) stores energy in the form of a magnetic field, generally given by $\frac{LI^2}{2}$, where L and I are inductance and operating

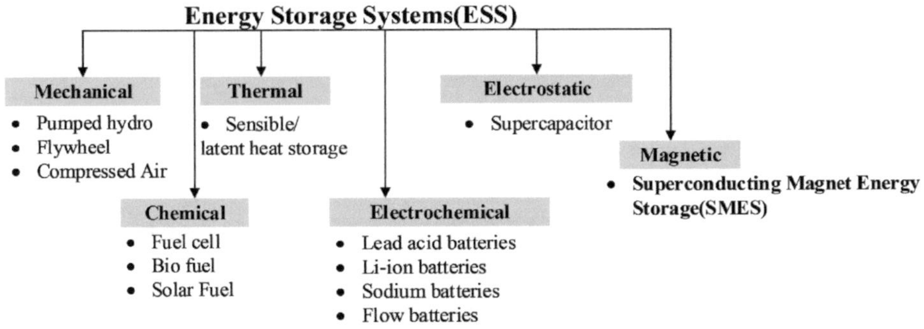

Figure 1.
Classification of various types of energy storage systems.

current, respectively. It is renowned for its exceptional efficiency, surpassing 95%, making it more efficient than any other Energy Storage System (ESS) available. The significant advantage of SMES is the quick charging and discharging capacity of energy in milliseconds, along with high energy density and a long life cycle.

2. Superconducting magnet energy storage (SMES)

SMES differs from other energy storage technologies, such as flywheels, batteries, and capacitors, which store energy in the form of potential energy. Meanwhile, SMES stores the energy in the form of kinetic energy. The electrical power system consists of generators, transmission lines, and motors, which are primarily inductive in nature. Hence, generation, transmission, and utilization are carried out using kinetic energy. Therefore, if energy storage is also in the form of kinetic energy, then there will be no energy conversion involved, which can increase the efficiency of electrical power storage and transfer. SMES is one such form of device that uses persistent and non-dissipative ($R_{DC} = 0$) properties of superconductors for electrical energy storage in the form of the magnetic field. The efficiency of the SMES system is said to be greater than 95%. The major applications of SMES are as follows:

1. *Power quality improvement*: SMES can improve power quality by means of power fluctuation compensation, dampening low frequency, instantaneous voltage sag compensation [1], power system stability, and load leveling. It can also be used as an energy storage device for Flexible AC Transmission System (FACTS) devices [2].

2. *Integration with renewable energy*: Renewable energy sources like solar [3] and wind power [4] can use SMES to store energy during surplus generation. The same stored energy can be used when the generation is low due to climate conditions.

3. *Pulsed power source for defense strategic applications*: Quick discharge of SMES can be used as a pulsed power supply for high energy pulsed laser and electromagnetic-based railgun and launch systems [5, 6].

4. *Uninterruptible Power Supplies (UPS)*: SMES can be used as a UPS for highly sensitive, high-performance devices where the quality of the power supply is very important [7].

5. *Electrical vehicle charging*: Fast charging stations of electric vehicles can utilize SMES for quick charging.

6. *Research and development*: SMES can be used in places where high energy is required in a very short duration.

2.1 Component of SMES

An SMES system, as shown in **Figure 2**, consists of a superconducting coil maintained at the cryogenic temperature inside a cryostat. The charging and discharging of SMES during operation is carried out using a Power Conversion System (PCS). The control unit determines the State Of Charge (SOC) of the SMES and maintains the temperature inside the cryostat.

2.1.1 Superconducting coil

A superconducting coil is the main component of a system in which energy is stored in the form of a magnetic field, which depends on current carrying capacity, which is a function of the shape of the magnet. The coil is a lossless inductor, and the stored energy is proportional to the square of the current and is given by Eq. 1

$$E = \frac{LI^2}{2} = \frac{1}{2\mu_o} \iiint_V B^2(x,y,z)\,dx\,dy\,dz \tag{1}$$

Where L is the inductance of the coil in Henry (H), I is the current in Ampere (A), and E is the stored energy in Joule (J). Since the energy can be said to be stored in the form of the magnetic field, thus the energy can be calculated using the volumetric sum of magnetic energy density $\left(\frac{B^2(x,y,z)}{2\mu_o}\right)$ in space surrounding the coil as given by the second expression. In a superconducting magnet, a high magnetic field can be

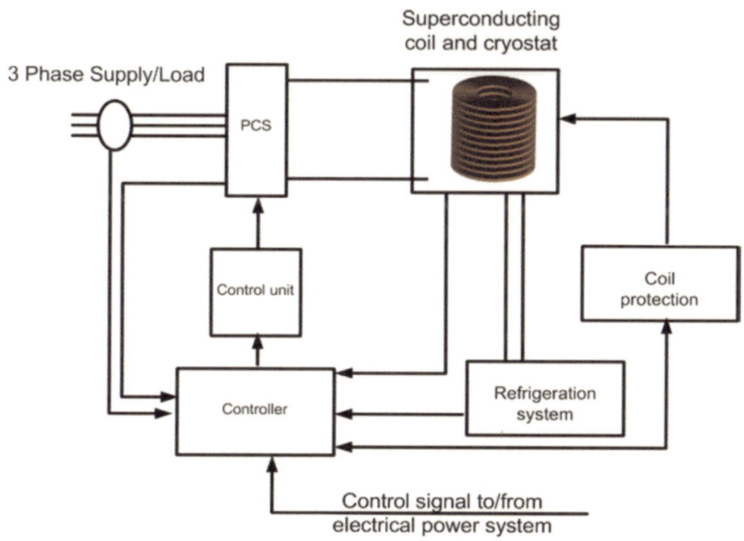

Figure 2.
Schematic of SMES system with major components.

generated thus stored energy can be very high depending upon the generated magnetic field.

The superconductor used to make the coil depends upon the operating condition and application of SMES. The coil can also be arranged in various configurations depending on the application. The coil is placed inside a vacuum Multi-Layer-Insulated (MLI) cryostat, which provides low temperature and protects it from heat in-leak from ambient and current leads during operation. A cryogenic refrigerator maintains the temperature inside the cryostat.

2.1.2 Configurations of SMES magnet coil

High-Temperature Superconducting (HTS) tape production is limited in length, typically not exceeding 1 km of continuous tape. Like Low-Temperature Superconducting (LTS) magnet coils, creating an HTS magnet coil with a continuous length would require numerous joints, which are resistive. These joints are not recommended as they can trigger the quench of the magnet coil. Therefore, HTS magnet coils are constructed by assembling individual modules or sections, each utilizing the available length of HTS tape.

When winding an HTS magnet coil, the only feasible configurations are the Single Pancake (SP) and Double Pancake (DP) coil due to the flat structure of HTS tape. The HTS tape-based SP and DP wound coil geometries are shown in **Figure 3(a)** and **(b)**, respectively. These individual coils can then be arranged in either a solenoid or toroidal configuration, as shown in **Figure 4(a)** and **(b)**.

Figure 5 displays the multi-solenoid configuration used in SMES, created by assembling individual solenoids based on their polarity. This configuration offers

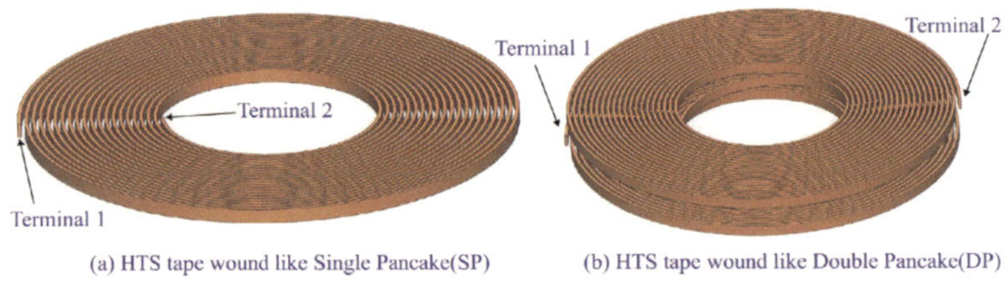

Figure 3.
Single pancake and double pancake coil winding.

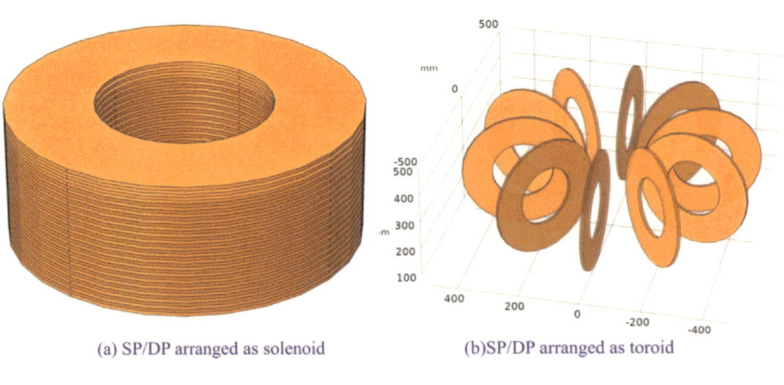

Figure 4.
Assembly of pancake coils as (a) solenoid and (b) toroid.

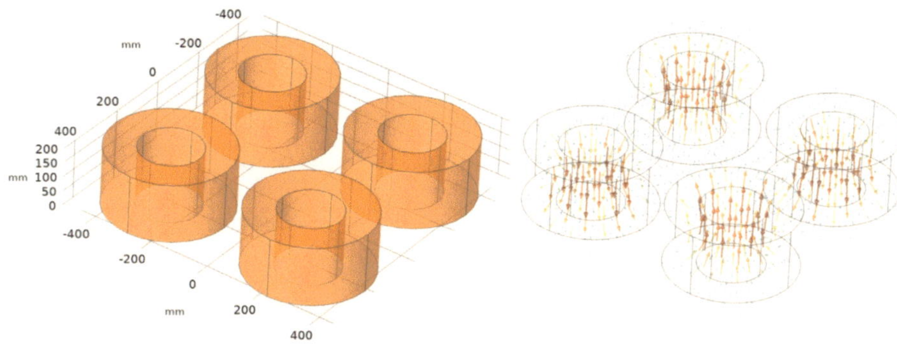

(a) Multisolenoid Configuration (b) Vector plot of Magnetic flux density

Figure 5.
Multipole solenoid configuration made by assembling solenoid.

several advantages, including reduced stray magnetic fields and lower stress levels compared to a single solenoid design. The reduction in stray magnetic fields is achieved by arranging adjacent coils with opposite polarities.

In our prior research, we optimized the dimensions of superconducting magnet coils using real-coded genetic algorithms and teaching-learning-based optimization techniques. The primary objective was to minimize the length of high-temperature superconducting (HTS) tape required to achieve a specified stored energy while ensuring that the stress remained within the critical limits. Additionally, we conducted shape optimization, resulting in novel configurations that require 7% less tape length compared to conventional solenoid structures, thereby reducing overall costs.

2.1.3 Refrigeration system

The superconducting coil must be below T_c to attain a superconducting state. A cryogenic refrigerator provides low temperatures to the system using a cryocooler-based conduction-cooled magnet (dry system) or as a recondenser for cryogen immersed magnet (wet system). The operating temperature of the SMES coil depends upon the type of superconductor used and its application (liquid helium for LTS and LN2 for HTS).

Liquid helium is also used for high-temperature superconductors for high magnetic field SMES. The power required for refrigeration increases as the operating temperature decreases. During the operation, if thermal stability breaks due to a rise in temperature because of heat in-leak from outside or generated in the coil, then refrigeration system capacity should also take care to avoid the quenching of the magnet.

2.1.4 Power conversion system

The PCS provides an interface between the SC magnet and the AC power grid. During surplus grid power, the superconducting magnet is charged with a DC power supply, which is provided by PCS with the help of rectifier and filter circuits. Whereas, during the dip in the power grid, the PCS automatically discharges the SC magnet and feeds the power to the grid with the help of a DC-link capacitor bank and grid-tie Voltage Source Inverters (VSI). Once the superconducting magnet is charged to its full capacity, it will be switched to the persistent mode in which the two terminals of the SC magnet will be shorted for the free-wheeling of current. To make PCS work in charging, persistent, and discharging modes, a DC chopper circuit is also incorporated within PCS.

Sub-subsections can also be used throughout the manuscript.

3. SMES around the globe

A superconducting magnet coil as an energy storage device was first proposed by N. Mohan in 1973 as a theoretical and economic study. A numerical study was performed for the performance of a superconducting magnet coil for power stability.

3.1 Low temperature superconductor SMES

The world's first SMES (30 MJ) was built by LAS atom laboratory, USA, and installed in Tacoma Substation of the Bonneville Power Administration in 1983 to demonstrate superconducting technology in the power utility sector [8, 9]. A 100 MJ LTS SMES was developed by Florida State University in 2003, operating at 4.5 K with liquid helium, using NbTi CICC for a solenoid-type coil with a 4 kA current and 10.8 H inductance [10]. Japan's first 7.3 MJ LTS SMES was developed in 2004 by Nagaya et al., using NbTi Rutherford cable with glass kapton insulation, operating at 2.6 kA and 2.08 H, cooled by liquid helium [11, 12]. In 2006, Japan developed a 2.9 MJ solenoid coil for advanced power control, using aluminum stabilized NbTi CICC with a 9.6 kA current [13]. KERI developed a 3 MJ LTS SMES in 2004 to improve power quality, with a solenoid magnet using NbTi Rutherford cable, operating at 1 kA and 6 H inductance [14, 15]. A 2.6 MJ LTS SMES was installed in the Elettra Synchrotron Light facility, Trieste-I Italy, to reduce voltage dips, with NbTi cable and HTS current leads, operating at 1100 A and 4.32 H [16]. Japan demonstrated a 270 kJ force-balanced toroid LTS SMES in 2005, using NbTi wire with a 552 A current and 1.8 H inductance [17–19]. Italy designed a 200 kJ LTS SMES in 2008, with an additional coaxial coil for a reduced stray field, operating at 150 A and 19 H inductance [20]. A 2 MJ multipole LTS solenoid SMES was tested by Wang et al. in 2010, with internal cooling channels, operating at 490 A and 16.644 H inductance [21]. Bhunia et al. designed a 4.5 MJ toroidal LTS SMES in 2014, using Rutherford-type NbTi with a 1200 A current [22, 23].

3.2 High temperature superconductor SMES

Kalsi et al. designed, fabricated, and tested the first solenoid HTS SMES (5 kJ) using 1G HTS wire (Bi-2223) in 1996 for EUS, Germany, to improve power quality in process plants [24]. The coils were cooled with a GM cryocooler operating at 100 A and 25 K, with a maximum central field of 0.97 T and an inductance of 1 H. Kurtz et al. designed a 150 kJ HTS SMES in 2003, operating at 20 K with BSCCO-2223 HTS tape, with a magnet coil cooled by a cryocooler and an operating current of 80 A [25]. Wang et al. fabricated a 30 kJ HTS SMES in 2008, protected by solid nitrogen for quenching, operating at 20 K and 155 A [26]. Under the DGA project, an 800 kJ HTS SMES was developed in 2005 using Bi2212 PIT tapes, operating at 20 K and 300 A [27–29].

The University of Wollongong fabricated a 2.5 kJ SMES in 2005 using BSCCO-2223 tape [30]. In 2006, a 1 MJ HTS SMES was installed in a live power grid suburb of Beijing, with a 16.4 km BSCCO HTS tape coil, operating at 564 A and 4.2 K, cooled by liquid helium and a cryocooler [31, 32]. Wojtasiewicz et al. designed a 7.3 kJ conduction-cooled HTS SMES, operating at 135 A and 35 K [33, 34]. KERI designed a 600 kJ HTS SMES in 2007 using 11 km of 4-ply BSCCO-2223 HTS tape, operating at 275 A and 20 K [35]. Chubu Electric and partners studied a 2 GJ toroid HTS SMES in 2012, with a 180 DP coil operating at 540 A and 20 K [36] . Kim et al. fabricated a 10 kJ toroid HTS SMES in 2011 using BSCCO wire, operating at 300 A and 6 K. Kozak et al. developed a 34.8 kJ HTS SMES in 2009, operating at 13 K and 264 A [37, 38] . Li et al. fabricated a 4 kJ Micro HTS SMES with a hybrid Li-ion battery system, operating at 30 K and 260 A [39]. Brookhaven National Laboratory built a 1.7 MJ HTS SMES coil in 2018, operating at 700 A and 4 K [40]. Julien et al. designed a 12 T High field HTS SMES in 2021 under the BOSS project, operating at 850 A and 4 K, with an inductance of 2.79 H [41].

Table 1 shows the development of SMES around the globe in chronological order.

Sl no.	Year	Country	Capacity	Superconductor used	Operating temp (K)	Coil configuration
1	1983	USA [8, 9, 42, 43]	30 MJ	NbTi	4.5 K	Solenoid
2	1997	USA [24]	5 kJ	Bi2223	25 K	Solenoid
3	2003	USA [10, 44]	100 MJ	NbTi	4.5 K	Solenoid
4	2003	Germany [25]	150 kJ	Bi2223	20 K	Solenoid
5	2004	Japan [11, 12]	7.34 MJ	NbTi	4 K	4 Pole coil
6	2005	Japan	1 MJ (90 KJ)	Bi2212	4.2 K	Solenoid
7	2005	Japan [13]	2.9 MJ	NbTi	5 K	Solenoid
8	2005	China [45, 46]	0.3 MJ	NbTi	4.2 k	Solenoid
9	2008	China [26]	30 kJ	Bi2223	10 K	Solenoid
10	2005	France [27–29]	800 kJ	Bi2222	20 K	Solenoid
11	2005	Australia [30]	2.48 kJ	Bi2223	25 K	Solenoid
12	2006	Japan [47, 48]	100 kJ	NbTi/Cu	4 K	Solenoid
13	2006	China [31, 32]	1 MJ	Bi2223	4.2 K	Solenoid
14	2006	Korea [14, 15]	3 MJ	NbTi	4.2 K	Solenoid
15	2006	Korea [35]	600 kJ	Bi2223	20 K	Solenoid
16	2006	Italy [16]	2.6 MJ	NbTi	4.2 K	Solenoid
17	2007	Japan	1 MJ	NbTi	4 K	Solenoid
18	2007	Japan [49]	6.5 MJ	NbTi, Bi2212	4.2 K	Solenoid
19	2007	China [50]	35 kJ	Bi2223	20 K	Solenoid
20	2008	Japan [36]	2.4 GJ	YBCO	20 K	Toriodal
21	2008	Japan [17–19]	270 kJ	NbTi	4.2 K	Toriodal

Sl no.	Year	Country	Capacity	Superconductor used	Operating temp (K)	Coil configuration
22	2008	Italy [20]	200 kJ	NbTi	4.2 K	Solenoid
23	2009	China [51]	10 kJ	Bi2223	10 K	
24	2009	Korea [52]	10 kJ	YBCO/BSCCO	20 K	Solenoid
25	2010	China [21]	2 MJ	NbTi	4.2 K	4 solenoid
26	2010	Korea [53]	2.5 MJ	Bi2223/YBCO	20 K	Toriodal
27	2010	Poland [37, 38]	34.8 kJ	Bi2223	12 K	Solenoid, DP
28	2014	India [22]	4.5 MJ	NbTi	4.2 K	Toriodal
29	2018	China [39]	4 kJ	REBCO	30 K	Solenoid, DP
30	2018	USA [40]	1.7 MJ	YBCO	4.2 K	Toriodal
31	2006	Poland [33, 34]	7.3 kJ	Bi-2223	35 K	Solenoid, DP
32	2019	China [54–56]	10 MJ	YBCO	20 K	Solenoid, DP
33	2022	France [41]	1 MJ	YBCO	4.2 K	Solenoid, DP

Table 1.
SMES around the globe.

4. Cooling methods used in SMES

4.1 Source of heat generation

The cooling method adopted and heat generation during the operation of SMES also affect the overall design of SMES. Each component of SMES, either cryostat, current lead, or the coil itself, is a potential source of heat during operation, and even a very low heat in-leak value can increase the temperature significantly at lower temperatures due to the very low heat capacity.

Major heat sources during operation SMES are:

1. Heat conduction through current leads from ambient to operating temperature (T_{Op})
2. Current lead joule heating due to high current flow
3. Ambient to operating temperature (T_{Op}) through cryostat
4. AC loss (Eddy current, Hysteresis and coupling) during charging and discharging
5. Joule heating in joints made of non-superconductor materials

Many factors are considered while designing the SMES. Most of the time, there is a need for compromise between various parameters, that is, higher operating current can increase heat in-leak in the system from copper current lead, and higher inductance can create very large voltage during charging and discharging, which can generate more eddy current leads to AC loss, magnetization loss in related nearby metallic component and even the insulation can breakdown in this high voltage.

4.2 Ways of achieving low temperature

Based on the operating temperature of SMES (4.2–30 K) mainly, three methods are being used.

1. Pool boiling using liquid helium

2. Conduction cooling using GM cryocooler

3. Solid nitrogen

Liquid helium is preferred for low LTS SMES $\left(T_{Op} = 4.2\text{ K}\right)$. SMES is cooled using the pool boiling method at the boiling temperature of liquid helium. This method requires constant replenishment of liquid helium or cryocooler to re-condense the evaporated helium back to the cryostat. For HTS, the preferred method is conduction cooling using a GM cryocooler, which significantly reduces the maintenance and operational hazard of liquid helium. Only one SC magnet for HTS SMES made recently [39] working operating at 30 K using solid nitrogen however, GM cryocooler is again used to solidify the nitrogen to attain 30 K.

4.3 Pool boiling using liquid helium

Helium has a boiling point of 4.2 K and is the only cryogenic available to provide a cryogenic temperature of SMES operating at 4.2 K.

Convection cooling (No phase change) and nucleate boiling are the only operating regimes for the cooling process. The maximum difference between surface temperature and LHe for peak nucleate boiling is approximately 1 K [57]. Factors affecting the peak heat flux(q^*) and ΔT_s significantly in the nucleate boiling region [58–60].

1. Temperature and pressure (subcooled [61] or supercritical)

2. Surface condition (roughness [62–65], coating Insulators [64], etc.)

3. Surface orientation (relative to gravitational force [65, 66])

4. Geometry (Channels [67], tubes [68])

5. Time to develop a steady state (transient heating [69–70])

A very necessary feature for an LTS SMES magnet is having a cooling channel [21] in between the layers. The cooling channel is provided by spacers in between the layers, as shown in **Figure 6**. Cooling channels provide uniform temperature in each part of the magnet, thus providing thermal stability to the system. In between two consecutive superconducting layers, an insulator layer is provided to avoid any sort of circuit between layers [71].

Earlier, LTS SMES had current lead made of copper, which was responsible for a very large heat leak, but it is now partially replaced by HTS lead. The HTS lead is cooler by evaporated helium vapor during operation.

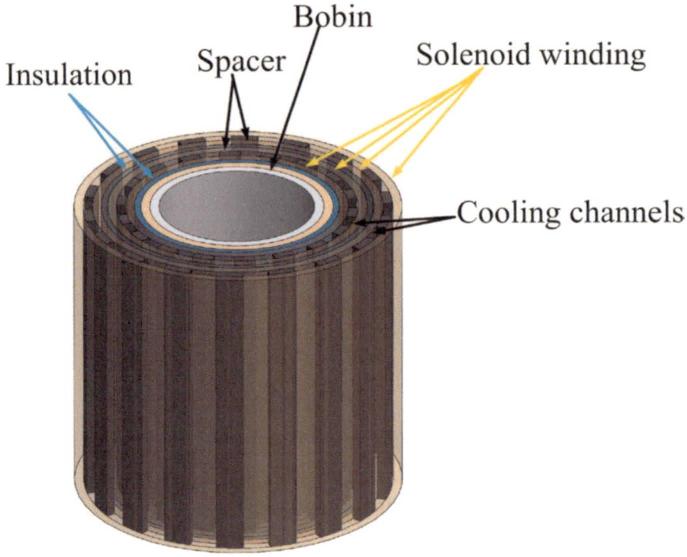

Figure 6.
LTS SMES showing cooling channels and spacers.

4.4 Conduction cooled

Conduction-cooled SMES is most preferred for coils made of HTS. This method eliminates the difficulties of handling liquid cryogens, thus simplifying maintenance and operation. The SMES coil magnet is mostly cooled by a single/double-stage GM cryocooler with an operating temperature range from 4.2 to 30 K.

4.4.1 GM cryocooler

GM cryocooler is based on the Gifford-McMahon cycle [72] proposed in 1959. The working fluid is mostly Helium. **Figure 7(a)** [73] and **7(b)** [74] show the commercially available single and double-stage GM cryocoolers, respectively. The temperature of the cold head also determines the cooling capacity. **Figure 8** shows the cooling capacity-temperature graph for CRYOMECH make single stage (AL640) GM cryocooler having 100 W cooling capacity at 20 K. **Figure 9** shows both 1st and 2nd stage working temperature along with the cooling capacity for NTE make 4 K double-stage cryocooler. Large cooling capacity GM cryocoolers are generally used for SMES applications.

Figure 10 shows the arrangement of major components of a conduction-cooled SMES. In all the conduction cooled, certain features are present to reduce the heat in-leak, thus providing stability during operation as listed below:

1. Radiation shield for magnet chamber

2. HTS current leads from copper lead to the coil

3. High vacuum inside chamber for low convection heat transfer

DOI: http://dx.doi.org/10.5772/intechopen.1008780

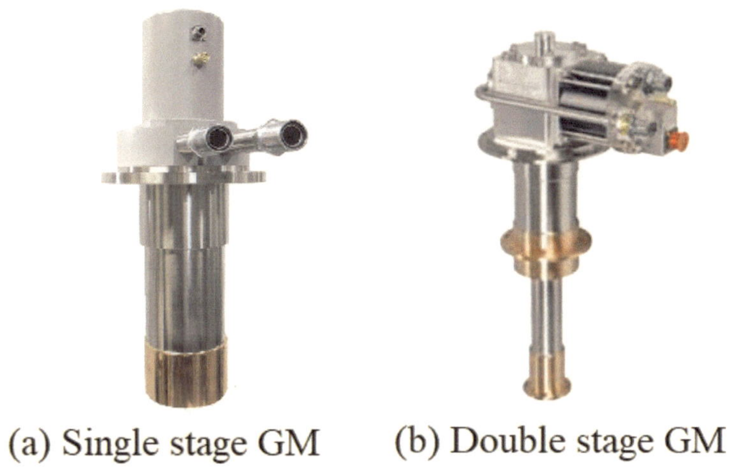

Figure 7.
Cryocooler by Cryomech (a) Single stage (b) Double stage.

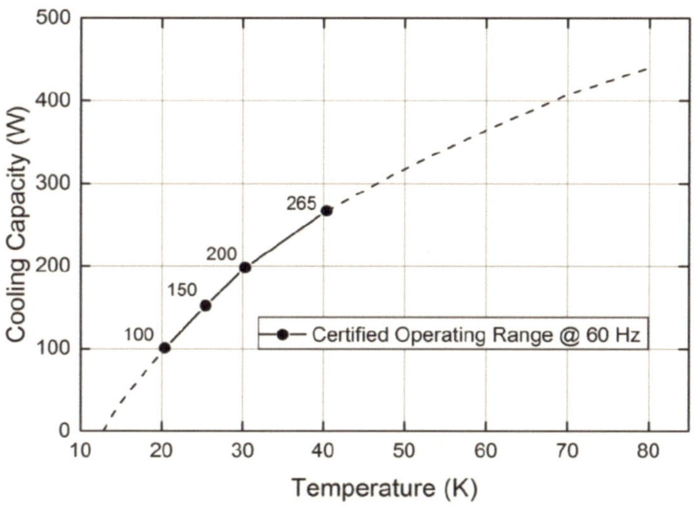

Figure 8.
Cooling capacity graph as per operating temperature for CRYOMECH AL630 [73].

4. Radiation shield for copper current lead from ambient temperature till thermal shield

5. Copper current lead cooled by 1st stage of cryocooler

6. Copper disk between pancakes connected with cryocooler 2nd stage through the copper thermal bridge.

7. Radiation baffles to reduce heat in-leak from the top surface.

Mainly, GM cryocooler has been used for conduction cooling for both magnet, current lead, and thermal shield, but a very large capacity HTS SMES (2.4 GJ Japan [36])

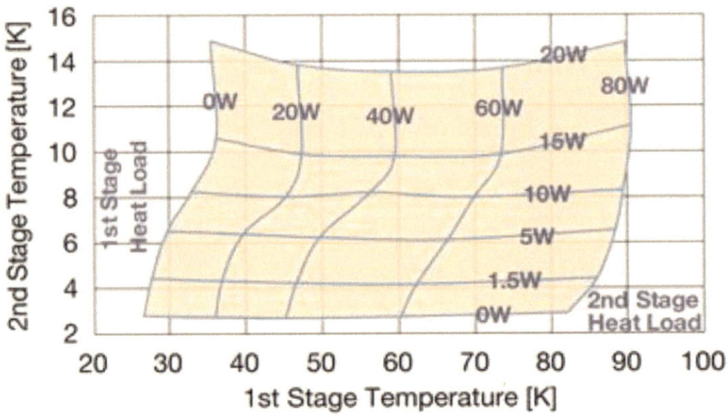

Figure 9.
Cooling capacity graph for both stages for different operating temperatures for NTE RDK-415D at 60 Hz [74].

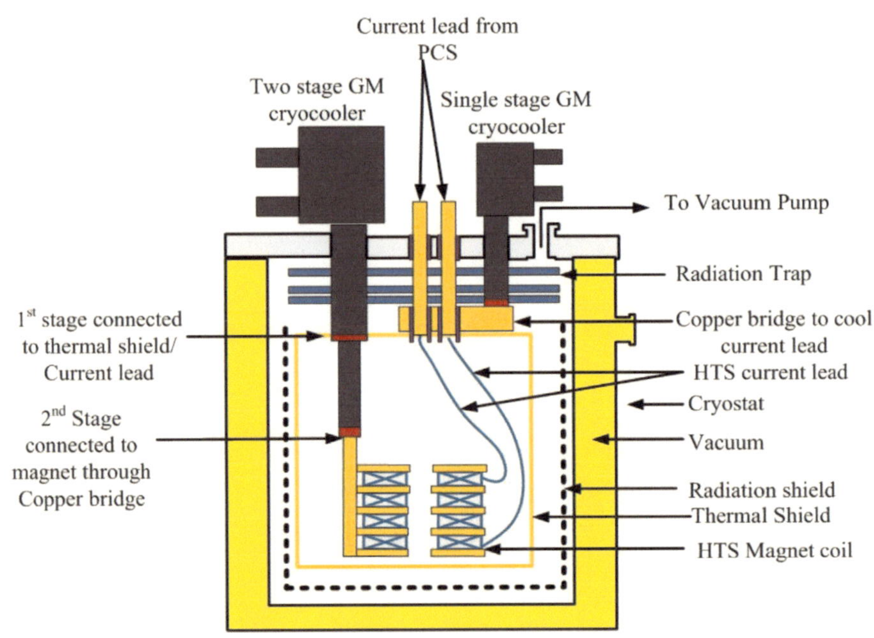

Figure 10.
Conduction cooled HTS SMES general trend.

used Stirling pulse tube cryocooler for current lead and thermal shield along with 94 GM cryocoolers for YBCO based coil.

Radiation shield property (emissivity) at cryogenic temperature and high vacuum depends strongly upon the surface conditions (polished, oxide layers), and thickness [75–77]. As radiation heat transfer involves electromagnetic waves [78], Parker and Abbot have given a relationship between emissivity, material resistivity, and absolute temperature [79].

Kawagoe et al. [47, 48] used spacers in conduction-cooled LTS SMES (100 kJ) made up of Dyneema FRP (DFRP) and litz wires connected to copper for improved conduction through indium foil.

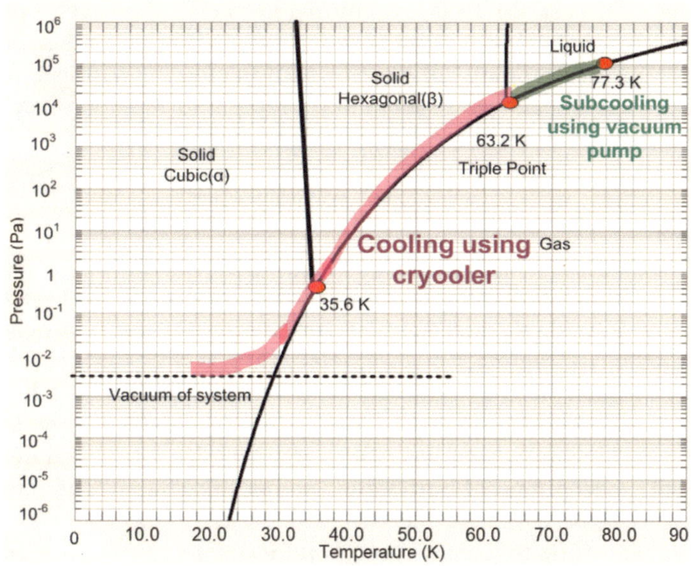

Figure 11.
Nitrogen phase diagram.

AIN sheets also used in between the cryocooler 2nd stage head and copper thermal bridge for insulation and keeping thermal conductivity.

4.5 Solid nitrogen

Solid nitrogen has a high heat capacity$\left(0.7 - 1.2J.cm^{-3}.K^{-1}\right)$ in the temperature range 20–30 K (six times more than copper) and a phase transition (cubic to hexagonal) at 35.6 K with the latent heat of 8.3 J/cm^{-3} [80] allowing it to keep the magnet safe if any thermal fluctuations occur. It has the only disadvantage that saturation vapor pressure of nitrogen facilitates higher convective heat in-leak during cooling and solidification. **Figure 11** shows the nitrogen phase diagram (Made by using Refprop [39, 81]) from a very low-pressure range to 10 times atmospheric pressure. The phase diagram also includes the path followed for getting solid nitrogen from the liquid. At first, the vacuum pump is utilized for subcooling to go from 77 to 63 K. After that, for solidification, the cryocooler is used to reduce further temperature.

Solid nitrogen has been used by Lee et al. [39] for better thermal stability of 4 kJ Hybrid HTS SMES (SMES coupled with Li-ion battery) during dynamic operations operating at 30 K [39]. At 30 K, the heat in-leak/generated can be easily compensated by cooling provided by GM cryocooler at higher wattage. The green and pink lines show the cooling curve on the phase diagram followed during the cool-down. To reduce the heat in-leak by convection using sold nitrogen, Lee et al. used a double-walled cryogenic vessel with two cryocoolers.

5. Technical challenges in SMES

Although there are many SMES projects have been carried out across the globe, it is still not yet commercialized. The major issue lies in the cryogenic technology and the cost of HTS tape. The cost of LTS is low compared to HTS; however, operating the LTS SMES at 4.2 K using liquid helium is expensive. With the great demand and advancement of 2G HTS tape manufacturing technology, the cost of HTS tape will likely reduce further. Quench detection and protection in superconducting magnets are another challenge that also needs to be addressed.

6. Conclusions

In conclusion, Superconducting Magnet Energy Storage (SMES) systems offer a highly efficient and rapid response solution for energy storage, significantly outperforming other technologies due to their ability to store and release energy with minimal losses. This chapter has explored the diverse applications and advantages of SMES, from improving power quality to integrating renewable energy sources, providing uninterruptible power supplies, and supporting high energy defense applications. The global development of both Low-Temperature Superconductor (LTS) and High-Temperature Superconductor (HTS) SMES systems highlights the advancements and ongoing efforts to harness superconducting technology for energy storage, showcasing innovations from the first 30 MJ SMES built in the USA to the recent 12 T High field HTS SMES designed in France.

However, despite significant progress, SMES technology faces several technical challenges that hinder its widespread commercialization. The primary issues revolve around the high costs associated with HTS tapes and the complex cryogenic technology required to maintain the superconducting state. The need for effective quench detection and protection mechanisms remains a critical area for further research and development. Nonetheless, with continuous advancements in HTS tape manufacturing and cryogenic systems, the future of SMES looks promising. By addressing these challenges, SMES can play a pivotal role in enhancing the efficiency and reliability of modern power systems, paving the way for a more sustainable and resilient energy infrastructure.

Author details

Ankit Anand[1*], Abhay Singh Gour[1], Tripti Sekhar Datta[1] and Vutukuru Vasudeva Rao[2]

1 Cryogenic Engineering Center, Indian Institute of Technology Kharagpur, Kharagpur, West Bengal, India

2 Indian Institute of Technology Tirupati, Tirupati, Andhra Pradesh, India

*Address all correspondence to: ankit5088@gmail.com

References

[1] Shikimachi K, Moriguchi H, Hirano N, Nagaya S, Ito T, Inagaki J, et al. Development of MVA class HTS SMES system for bridging instantaneous voltage dips. IEEE Transactions on Applied Superconductivity. 2005;**15**(2 PART II):1931-1934

[2] Mitani Y, Tsuji K, Murakami Y. Application of superconducting magnet energy storage to improve power system dynamic performance. IEEE Transactions on Power Apparatus and Systems. 1988;**3**(4):1418-1425

[3] Salama HS, Vokony I. Comparison of different electric vehicle integration approaches in presence of photovoltaic and superconducting magnetic energy storage systems. Journal of Cleaner Production. 2020;**260**:121099

[4] Shi J, Tang YJ, Ren L, Li JD, Chen SJ. Application of SMES in wind farm to improve voltage stability. Physica C: Superconductivity and its Applications. 2008;**468**(15):2100-2103. Proceedings of the 20th International Symposium on Superconductivity (ISS 2007)

[5] Bhardwaj A, Moon J, Pamidi SV. Design of power conversion system for pulsed power applications of SMES. In: Southeast Conferrence 2024. USA: IEEE; 2024. pp. 1464-1469

[6] Juengst KP, Salbert H. Fast SMES for generation of high power pulses. IEEE Transactions on Magnetics. 1996;**32**(4): 2272-2275

[7] Luo X, Wang J, Dooner M, Clarke J. Overview of current development in electrical energy storage technologies and the application potential in power system operation. Applied Energy. 2015; **137**:511-536

[8] Rogers JD, Schermer RI, Miller BL, Hauer JF. 30-MJ superconducting magnetic energy storage system for electric utility transmission stabilization. Proceedings of the IEEE. 1983;**71**(9): 1099-1107

[9] Boenig HJ, Hauer JF. Commissioning tests of the Bonneville power administration 30 MJ superconducting magnetic energy storage unit. IEEE Power Engineering Review. 1985;**PER-5** (2):32-33

[10] Luongo CA, Baldwin T, Ribeiro P, Weber CM. A 100 MJ SMES demonstration at FSU-CAPS. IEEE Transactions on Applied Superconductivity. 2003;**13**(2 II): 1800-1805

[11] Nagaya S, Hirano N, Kondo M, Tanaka T, Nakabayashi H, Shikimachi K, et al. Development and performance results of 5 MVA SMES for bridging instantaneous voltage dips. IEEE Transactions on Applied Superconductivity. 2004;**14**(2):699-704

[12] Nagaya S, Hirano N, Moriguchi H, Shikimachi K, Nakabayashi H, Hanai S, et al. Field test results of the 5 MVA SMES system for bridging instantaneous voltage dips. IEEE Transactions on Applied Superconductivity. 2006;**16**(2): 632-635

[13] Hayashi H, Hatabe Y, Nagafuchi T, Taguchi A, Terazono K, Ishii T, et al. Test results of power system control by experimental SMES. IEEE Transactions on Applied Superconductivity. 2006;**16** (2):598-601

[14] Cho JW, Kim HJ, Bae JH, Sim KD, Seong KC. Fabrication and test of a 3 MJ SMES magnet. IEEE Transactions on Applied Superconductivity. 2004;**14**(2): 743-745

[15] Kim HJ, Seong KC, Cho JW, Bae JH, Sim KD, Kim S, et al. 3 MJ/750 kVA SMES system for improving power quality. IEEE Transactions on Applied Superconductivity. 2006;**16**(2):574-577

[16] Ottonello L, Canepa G, Albertelli P, Picco E, Florio A, Masciarelli G, et al. The largest Italian SMES. IEEE Transactions on Applied Superconductivity. 2006;**16**(2):602-607

[17] Nomura S, Tsutsui H, Watanabe N, Suzuki C, Kajita S, Ohata Y, et al. Demonstration of the stress-minimized force-balanced coil concept for SMES. IEEE Transactions on Applied Superconductivity. 2003;**13**(2 II): 1852-1855

[18] Nomura S, Ohata Y, Hagita T, Tsutsui H, Tsuji-Iio S, Shimada R. Construction of a 7-T force-balanced helical coil. IEEE Transactions on Applied Superconductivity. 2005;**15**(2 PART II):1911-1914

[19] Nomura S, Kasuya K, Tanaka N, Tsuboi K, Tsutsui H, Tsuji-Iio S, et al. Quench properties of a 7-T force-balanced helical coil for large-scale SMES. IEEE Transactions on Applied Superconductivity. 2009;**19**(3): 2004-2007

[20] Morandi A, Breschi M, Fabbri M, Negrini F, Penco R, Perrella M, et al. Design, manufacturing and preliminary tests of a conduction cooled 200 kJ Nb-Ti μSMES. IEEE Transactions on Applied Superconductivity. 2016;**18**(2):697-700

[21] Wang Q, Dai Y, Zhao B, Song S, Cao Z, Chen S, et al. Development of large scale superconducting magnet with very small stray magnetic field for 2 MJ SMES. IEEE Transactions on Applied Superconductivity. 2010;**20**(3):1352-1355

[22] Bhunia U, Akhter J, Nandi C, Pal G, Saha S. Design of a 4.5 MJ/1 MW sectored toroidal superconducting energy storage magnet. Cryogenics. 2014;**63**:186-198

[23] Bhunia U, Saha S, Chakrabarti A. Pareto optimal design of sectored toroidal superconducting magnet for SMES. Physica C: Superconductivity and its Applications. 2014;**505**:6-13

[24] Kalsi SS, Aized D, Connor B, Snitchler G, Campbell J, Schwall RE, et al. HTS SMES magnet design and test results. IEEE Transactions on Applied Superconductivity. 1997;**7**(2 PART 1): 971-976

[25] Kreutz R, Salbert H, Krischel D, Hobl A, Radermacher C, Blacha N, et al. Design of a 150 kJ high-Tc SMES (HSMES) for a 20 kVA uninterruptible power supply system. IEEE Transactions on Applied Superconductivity. 2003;**13** (2 II):1860-1862

[26] Wang Q, Dai Y, Song S, Wen H, Bai Y, Yan L, et al. A 30 kJ Bi2223 high temperature superconducting magnet for SMES with solid-nitrogen protection. IEEE Transactions on Applied Superconductivity. 2008;**18**(2):754-757

[27] Tixador P, Bellin B, Deleglise M, Vallier JC, Bruzek CE, Allais A, et al. Design and first tests of a 800 kJ HTS SMES. IEEE Transactions on Applied Superconductivity. 2007;**17**(2):1967-1972

[28] Bellin B, Tixador P, Deleglise M, Vallier JC, Pavard S, Bruzek CE. Cryogenic design of a 800 kJ HTS SMES. Journal of Physics: Conference Series. 2006;**43**(1):817-820

[29] Tixador P, Deleglise M, Badel A, Berger K, Bellin B, Vallier JC, et al. First tests of a 800 kJ HTS SMES. IEEE Transactions on Applied Superconductivity. 2008;**18**(2):774-778

[30] Hawley CJ, Gower SA. Design and preliminary results of a prototype HTS SMES device. IEEE Transactions on Applied Superconductivity. 2005;**15**(2 PART II):1899-1902

[31] Dai S, Xiao L, Wang Z, Zhang J, Zhang D, Hui D, et al. Design of a 1 MJ/0.5 MVA HTS magnet for SMES. IEEE Transactions on Applied Superconductivity. 2007;**17**(2):1977-1980

[32] Xiao L, Wang Z, Dai S, Zhang J, Zhang D, Gao Z, et al. Fabrication and tests of a 1 MJ HTS magnet for SMES. IEEE Transactions on Applied Superconductivity. 2008;**18**(2):770-773

[33] Wojtasiewicz G, Janowski T, Kozak S, Kondratowicz-Kucewicz B, Kozak J, Surdacki P, et al. HTS magnet for 7.3 kJ SMES system. Journal of Physics: Conference Series. 2006;**43**(1):821-824

[34] Wojtasiewicz G, Janowski T, Kozak S, Kondratowicz-Kucewicz B, Majka M. Bi-2223 magnet for the model of SMES. Journal of Physics: Conference Series. 2008;**97**:012019

[35] Kim WS, Kwak SY, Lee JK, Choi KD, Jung HK, Seong KC, et al. Design of HTS magnets for a 600 kJ SMES. IEEE Transactions on Applied Superconductivity. 2006;**16**(2):620-623

[36] Shikimachi K, Hirano N, Nagaya S, Kawashima H, Higashikawa K, Nakamura T. System coordination of 2 GJ class YBCO SMES for power system control. IEEE Transactions on Applied Superconductivity. 2009;**19**(3):2012-2018

[37] Kozak J, Kozak S, Janowski T, Majka M. Design and performance results of first polish SMES. IEEE Transactions on Applied Superconductivity. 2009;**19**(3):1981-1984

[38] Kozak J, Majka M, Kozak S, Janowski T. Performance of SMES system with HTS magnet. IEEE Transactions on Applied Superconductivity. 2010;**20**(3):1348-1351

[39] Li Y, Song P, Kang Y, Feng F, Qu T. Design of a 30-K/4-kJ HTS magnet cryocooled with solid nitrogen. IEEE Transactions on Applied Superconductivity. 2018;**28**(4):1-6

[40] Gupta R, Anerella M, Joshi P, Higgins J, Lalitha S, Sampson W, et al. Design, construction, and testing of a large-aperture high-field HTS SMES coil. IEEE Transactions on Applied Superconductivity. 2016;**26**(4):1-8

[41] Vialle J, Badel A, Tixador P. 12 T insulated REBCO magnet used as 1 MJ SMES: Protection strategies and preliminary assembly tests. IEEE Transactions on Applied Superconductivity. 2022;**32**(6):1-7

[42] Miller BL. 30 MJ SMES BPA'S Tacoma substation engineering. In: Proceedings of the DOE Physical and Chemical Energy Storage Annual Contractors' Review Meeting; 23-26 August 1982; Stouffers National Center Hotel, Arlington, Virginia. The Department; 1982. pp. 197-200

[43] Hauer JF. Utility controls for BPA 30-MJ SMES system. In: Proceedings of the DOE Physical and Chemical Energy Storage Annual Contractor's Review Meeting. Springfield, Virginia: National Technical Information Service, NTIS, U. S. Department of Commerce; 1982. pp. 201-205

[44] Arsoy AB, Wang Z, Liu Y, Ribeiro PF. Transient modeling and simulation of a SMES coil and the power electronics interface. IEEE Transactions on Applied Superconductivity. 1999;**9**(4):4715-4724

[45] Jiang X, Zhu X, Cheng Z, Ren X, He Y. A 150 kVA/0.3 MJ SMES voltage sag compensation system. IEEE Transactions on Applied Superconductivity. 2005;**15** (2 PART II):1903-1906

[46] Energy Stored Coil A. A 0.3 MJ SMES magnet of a voltage sag. IEEE Transactions on Applied Superconductivity. 2004;**14**(2):717-720

[47] Kawagoe A, Yamamuro H, Sumiyoshi F, Mito T, Chikaraishi H, Baba T, et al. Heat transfer properties of a conduction cooled prototype LTS pulse coil for UPS-SMES. IEEE Transactions on Applied Superconductivity. 2006;**16** (2):624-627

[48] Mito T, Kawagoe A, Chikaraishi H, Maekawa R, Okumura K, Abe R, et al. Validation of the high performance conduction-cooled prototype LTS pulse coil for UPS-SMES. IEEE Transactions on Applied Superconductivity. 2006;**16** (2):608-611

[49] Tosaka T, Koyanagi K, Ohsemochi K, Takahashi M, Ishii Y, Ono M, et al. Excitation tests of prototype HTS coil with Bi2212 cables for development of high energy density SMES. IEEE Transactions on Applied Superconductivity. 2007;**17**(2):2010-2013

[50] Shi J, Tang Y, Zhou Y, Chen J, Xu D, Wang H, et al. Development of a conduction-cooled HTS SMES. IEEE Transactions on Applied Superconductivity. 2007;**17**(3):3846-3851

[51] Chen X, Jin J, Xin Y, Zhang J. Design and optimization of HTS magnets for SMES applications. 2009 International Conference on Applied Superconductivity and Electromagnetic Devices, ASEMD. 2009;**2009**:26-29

[52] Kim AR, Jung HY, Kim JH, Park M, Yu IK, Kim SH, et al. Design of a 10 kJ class SMES model coil for real time digital simulator based grid connection study. IEEE Transactions on Applied Superconductivity. 2009;**19**(3):2023-2027

[53] Choi JH, Choi JW, Kim HJ, Seong K, Lee S, Kim SH. Basic insulation characteristics of 2.5 MJ class his SMES. IEEE Transactions on Applied Superconductivity. 2009;**19**(3):2040-2043

[54] Ji K, Long W, He J, Zheng J. Analysis of Eddy-current loss and its reduction in a conduction-cooled HTS magnet. IOP Conference Series: Earth and Environmental Science. 2019;**300**(4):1-8

[55] Peng SS, Zheng J, Li WY, Dai YJ. AC loss analysis of a single-solenoid HTS SMES based on H-formulation. IOP Conference Series: Earth and Environmental Science. 2019;**233**(2):1-7

[56] Zheng J, Peng SS, Li WY, Dai YJ. Magnet design of 10MJ multiple solenoids SMES. IOP Conference Series: Earth and Environmental Science. 2019; **233**(3):1-8

[57] James GB, Lewis KG, Maddock BJ. Critical heat fluxes for liquid helium boiling in small channels. Cryogenics. 1970;**10**(6):480-485

[58] Van Sciver SW. Classical Helium Heat Transfer. New York: Springer; 2012. pp. 115-161

[59] Cummings R, Smith J. Boiling Heat Transfer to Liquid Helium. United Kingdom: Pergamon Press Ltd; 1966

[60] Bald W, Wang TY. The nucleate pool boiling dilemma. Cryogenics. 1976; **16**(5):314-315

[61] Ibrahim E, Boom R, McIntosh G. Heat transfer to subcooled liquid helium. In: Advances in Cryogenic Engineering. Boston, MA: Springer; 1978. pp. 333-339

[62] Boissin J, Thibault J, Roussel J, Faddi E. Boiling heat transfer and peak nucleate boiling flux in liquid helium. In: Advances in Cryogenic Engineering. Boston, MA: Springer; 1995. pp. 607-616

[63] Ogata H et al. Heat Transfer to Boiling Helium from Machined and Chemically Treated Copper Surfaces. New York, NY: Springer; 1982

[64] Granieri PP, Van Weelderen R. Deduction of steady-state cable quench limits for various electrical insulation schemes with application to LHC and HL-LHC magnets. IEEE Transactions on Applied Superconductivity. 2014;**24** (3):8-13

[65] Lyon D. Boiling heat transfer and peak nucleate boiling fluxes in saturated liquid helium between the ramda and critical temperatures. Advances in Cryogenic Engineering. 1965;**10**:371

[66] Chen Z, Van Sciver S. Channel heat transfer in he i-steady state orientation dependence. In: Advances in Cryogenic Engineering. Springer; 1986. pp. 431-438

[67] Wilson M. Heat Transfer to Boiling Liquid Helium in Narrow Vertical Channels. United Kingdom: Pergamon Press Ltd; 1966

[68] Lehangre J, Boissin C, Johannes C, de La Harpe A. Critical Nucleate Boiling of Liquid Helium in Narrow Tubes and Annuli. In: Proceedings of the 2nd International Cryogenics Engineering Conference. Brighton: Hiffe Science and Technology; 1968. p. 274

[69] Giarratano PJ, Frederick NV. Transient Pool Boiling of Liquid Helium Using a Temperature-Controlled Heater Surface. US, Boston, MA: Springer; 1980. pp. 455-466

[70] Cowley CW, Timson WJ, Sawdye JA. A method for improving heat transfer to a cryogenic fluid. In: Timmerhaus KD, editor. Advances in Cryogenic Engineering. Boston, MA: Springer US; 1962. pp. 385-390

[71] Wyatt AFG. Kapitza Conductance of Solid-Liquid He Interfaces. Boston, MA: Springer US; 1981. pp. 31-72

[72] Gifford WE. The Gifford-McMahon cycle. In: Timmerhaus KD, editor. Advances in Cryogenic Engineering. Boston, MA: Springer US; 1966. pp. 152-159

[73] Bluefors. Gifford-McMahon Cryocoolers. n.d. Available from: https://bluefors.com/products/gifford-mcmahon-cryocoolers/

[74] Nagase NTE. RDK-415D [Cold Refrigerator]. n.d. Available from: https://www.nagase-nte.co.jp/product/cold/cold-RDK-415D.html

[75] Hawks KH, Cottingham WB. Total Normal Emittances of some Real Surfaces at Cryogenic Temperatures. Boston, MA: Springer US; 1971. pp. 467-474

[76] Obert W, Copland JR, Hammond DP, Cook T, Harwood K. Emissivity Measurements of Metallic Surfaces Used in Cryogenic Applications. New York: Plenum Press, Boston, MA; 1982. pp. 293-300

[77] Ruccia FE, Hinckley RB. The surface emittance of vacuum-metallized polyester film. In: Timmerhaus KD, editor. Advances in Cryogenic Engineering. Boston, MA: Springer US; 1967. pp. 300-307

[78] Barron RF, Nellis GF. Cryogenic Heat Transfer. Boca Raton: CRC Press; 2017

[79] Parker W, Abbott G. Theoretical and experimental studies of the total

emittance of metals. In: Symposium on Thermal Radiation of Solids. Washington, DC: US Government Printing Office; 1965. pp. 11-28

[80] YI. (auth.). Case Studies in Superconducting Magnets: Design and Operational Issues. 2nd ed. Springer US; 2009

[81] Lemmon EW, Bell IH, Huber ML, McLinden MO. NIST Standard Reference Database 23: Reference Fluid Thermodynamic and Transport Properties-REFPROP, Version 10.0. Gaithersburg: National Institute of Standards and Technology, Standard Reference Data Program; 2018. DOI: 10.18434/T4/1502528

Chapter 6

Low-Temperature Performance and Manufacture of Metals

Wangjun Cheng

Abstract

In this chapter, the background and significance of low-temperature manufacture and performance of metals were presented. The applications of low-temperature technology in typical fields, such as electronics, healthcare, aerospace, food engineering, new energy, and cryogenic manufacturing, were presented. Principles of cryogenic cutting, cryogenic casting, cryogenic treatment, and cryogenic forming in a low temperature were highlighted. Likewise, mechanical properties and hardening behaviors of metals, such as aluminum alloys and stainless steels at cryogenic temperatures, were discussed. Macroscopic and microscopic mechanisms of the above deformed metals at ultra-low temperatures were also revealed. Finally, cryogenic forming methods and properties were discussed. Cryogenic forming products and potential applications were demonstrated.

Keywords: low-temperature manufacture, cryogenic cutting, cryogenic casting, cryogenic treatment, cryogenic forming, performance

1. Introduction

The important embodiment of new quality productivity in the field of refrigeration is the development of original innovation capabilities, breakthroughs in key technologies, and the shaping of new tracks in traditional metal manufacturing. Low-temperature technology refers to the use of artificial refrigeration to achieve and maintain the desired cryogenic state of an object or a space. Generally, the temperature from ambient temperature to 123 K (−150°C) is called the universal cooled zone. From 123 K (−150°C) to 0 K (−273°C) is called the deep cooled zone. Low-temperature technology has broad application prospects in industrial and military production fields.

In the electronic manufacturing field, low-temperature technology is needed to control the production and testing temperature of chips and further ensure performance and stability [1]. Low-temperature technology can also improve production efficiency and product quality in chemical and pharmaceutical industries. In the aerospace field, low-temperature technology can adjust the temperature inside satellites to ensure their normal operation. In the new energy field, low-temperature technology can be used to regulate the temperature of solar photovoltaic panels, control the temperature of wind power generation systems, and improve the stability and reliability of the system.

The refrigeration industry has developed rapidly in the field of military and civil equipments with the progress of science and technology in today's world [2]. Further, the rapid development of refrigeration has promoted the emergence of low-temperature manufacturing technologies. In the field of electronic product manufacturing, certain substances at cryogenic temperatures embody completely different physical properties from those at ambient temperature. For example, cryogenics is needed to control the production and test the temperature of chips to ensure the performance and its stability. Cryogenics is also utilized to design cooled electronic devices, such as infrared detectors, quantum story enlargers, parametric amplifiers, and optical devices [3]. The codetection rate is greatly improved under the operating temperature from 77 to 42 K. Likewise, some small coolers have the advantages of high reliability, long life, low power consumption, and fast cooling.

Product quality in the chemical and pharmaceutical industries can be improved by low-temperature technology. The use of cryogenic temperatures allows the cells in the lesion to be killed and promotes the healed muscle. Cell death is not caused by freezing itself. However, the concentration of the solution in the cell is altered by the freezing temperature and rate. Further, the environment for bacterial production is altered and death occurs. As a result, a series of cryogenic medical machines and medical devices has been created. At present, electronic refrigerator, electronic freezing gun, electronic cryogenic cooler, electronic thermostat, cold needles, cold knife, cooling cap and cold therapy, cryostasis, cryomedicine, cryosurgery to remove cataracts, cultivation of IVF, etc. have been used in clinical practice. And cryotherapy machines above −100°C, cryotherapy machines at −100°C, and superconducting devices have been currently in service in the treatment of cancerous tumors.

The temperature inside a satellite can be regulated by low-temperature technology to ensure its proper functions in aerospace. Cryogenic fuels can be used as propellants in chemical rockets or solar energy. And liquid nitrogen can be used to precool and clean cryogenic propellant rockets. It can also be used for temperature control of radar microwave components, the cooling of infrared sensors, cooling of optical microscopes, etc. Rocket components with complex shapes can be produced by utilizing low-temperature stamping technology. Presently, ultra-thin, ultra-fine, and ultra-wide aluminum alloy materials can be successfully prepared.

The freshness period of food can be significantly extended by low-temperature technology in food engineering. The quality, color, taste, and freshness of meat, fresh shrimp, fruits, and vegetables are enhanced via low-temperature technology. The most common storage method in the food industry is low-temperature storage. It maintains the original quality of the food to a large extent, and hence the shelf life is extended. Common storage methods at low temperatures can be divided into cold storage (0 ~ 4°C), ice temperature (−0.5 ~ −2.8°C), microfrozen (−4°C), and frozen (−12 ~ −18°C). Low-temperature technology can also be used for the production and transportation in food engineering, such as the freeze-concentration technology, freeze-drying technology, freeze-milling technology, vacuum freeze-drying technology, etc. [4]. Low-temperature technology can also be used to dry various types of food. Utilizing the principle of sublimation of ice crystals, the moisture in the water-containing material is first frozen into ice. Then the ice in the material is directly sublimated into water vapor, and finally the material is dehydrated.

Low-temperature technology can be used to regulate the temperature of solar photovoltaic panels in the new energy field. The temperature of wind power systems can be controlled. The stability and reliability of the system are improved. In addition, cryogenic technology in the field of natural gas clean energy has also made a remarkable progress. For example, LNG was produced from coke oven gas at low temperatures, and the recovery rate of LNG is improved. Energy consumption was reduced. The quality of the product is improved. The application of low-temperature technology for hydrogen fuel cells has made it possible to realize nondestructive cold-starting even under the extreme conditions of −40°C. It is also possible to use low-temperature technology to improve the performance of hydrogen fuel cells and to improve the efficiency of hydrogen fuel cells [5].

Products can be processed by utilizing the special properties of substances in a cryogenic state. This technique is particularly useful in the processing of materials, such as aluminum alloys, copper alloys, and titanium alloys. These materials are difficult to process and shape at room temperature, but it is easy to process at ultra-low temperatures. The main technologies include cryogenic cutting, cryogenic casting, cryogenic treatment, cryogenic EDM, cryogenic grinding, and low-temperature machining. Also, low-temperature manufacturing has significant advantages for increasing the plasticity of materials and for improving their properties [6].

The applications of low-temperature technology in typical fields such as electronics, healthcare, aerospace, food engineering, new energy, and cryogenic manufacturing were presented in this paper. Principles of cryogenic cutting, cryogenic casting, cryogenic treatment, and cryogenic pressure processing in ultra-low-temperature deep-cooling manufacturing were highlighted. Meanwhile, mechanical properties and hardening behaviors of metals, such as aluminum alloys and stainless steels in low-temperature deformations, were discussed. Macroscopic and microscopic deformation mechanisms of the above metals at ultra-low temperatures were revealed. Further, cryogenic forming processes and properties were discussed. Finally, cryogenic forming products and potential applications were also demonstrated.

2. Low-temperature cutting

Low-temperature cutting is usually a cutting method that uses liquid nitrogen (LN_2) or carbon dioxide to keep the workpiece, tools, or cutting area in a low-temperature cooling state. To improve cutting efficiency, the cooling process is conducted by spraying the coolant or directly immersing workpiece in a low-temperature medium. This allows the cutting capacity of alloys to be increased and reduced tool wear due to the rapid heat dissipation during the cutting process.

According to various cooling sources, it can be divided into cryogenic liquid or gas cooling cutting, electronic cooling cutting, tool internal cooling cutting, and chemical cooling cutting [7]. The cutting heat is generated by the rapid friction and wear from the workpiece and tools. Further, the workpiece and tool are simultaneously heated, and the temperature can reach several hundred degrees centigrade. The energy in the metal cutting can be mainly divided into the cutting work, friction work, deformation work of surface layer, and surface energy. Due to the severe shear deformation, most of this work becomes heat. A residual 10% work remains in reserve in the cut alloy. Most of the friction work becomes heat. More than 90% of the total cutting work is converted to heat. Additionally, the cutting heat is mainly concentrated near the tip of the tool. This leads to an increase in the temperature of the tool and the workpiece [8]. Therefore, the cooling effect can be achieved by pouring coolant to reduce the cutting temperature. Likewise, the frictional resistance of the front and back of tools can be reduced by the coolant lubrication effect. The generated heat is reduced, and then the wear of the tool tip is reduced. To illustrate the principles of cryogenic cutting, examples of coolant spraying cutting and coolant immersing cutting are shown in Sections 2.1 and 2.2.

2.1 Coolant spraying cutting

This method allows a coolant (cryogenic gas or liquid) to spray onto the surface of the machined part and leads to a quickly achieved cooling effect of the cutting position and tools [9]. The experimental device for the coolant spraying cutting is shown in **Figure 1**. The low-pressure gas or liquid is applied as the refrigerant. After the expansion of the gas and liquid, the compressed air is ejected into the vortex chamber to form a free vortex. The internal gas is separated into the cold and hot gas flow after the exchange of kinetic energy. And then a vortex is generated in the tube. Also, the cooling air outlet is connected to the cutting oil to improve the cooling effect. The cutting oil is brought in by an airflow ejection to form a spray-like cooling oil with an easily achieved cooling of the workpiece. This hybrid cooling method is better than the pure gas cooling. It is favorable to the chip dispersion and does not pollute the environment. However, this method belongs to local external cooling, and the cooling zone is uneven and ineffective. Therefore, it is only suitable for the cutting of shaft parts and long pipe parts.

2.2 Coolant immersing cutting

A coolant immersing cutting method is proposed to solve the problem of uneven cooling during the low-temperature cutting. The workpiece is immersed in the freezing liquid to make the freezing temperature of the workpiece uniform and consistent, as seen in **Figure 2**. This device is equipped with a cooling box on a workbench and a mounted caliper. The workpiece during the cutting is immersed in a coolant (alcohol and LN_2). Then the workpiece is machined with only the surface exposed for the

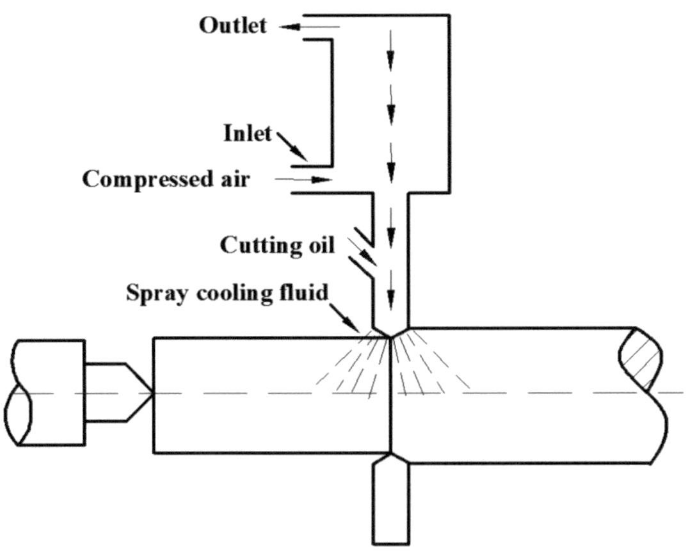

Figure 1.
Experimental principle for coolant spraying cutting.

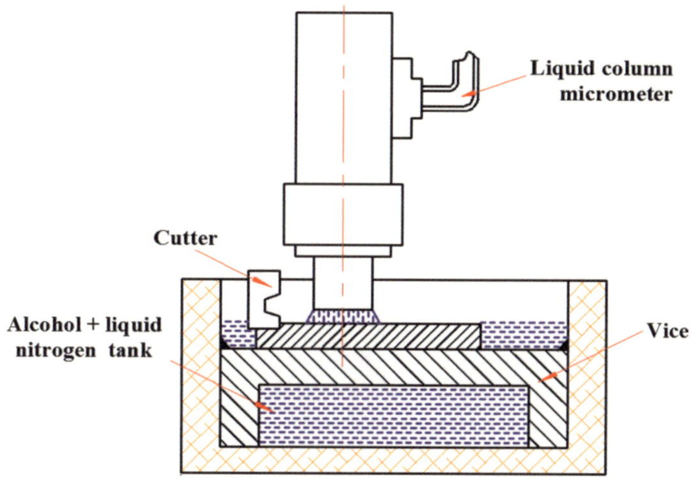

Figure 2.
Experimental principle for coolant immersing cutting.

cutting. This method is suitable for grinding, planning, drilling, milling, broaching, and other cutting methods. However, it causes environmental pollution, requires a large consumption of coolant, and has an inability to process rotating surfaces. Additionally, it is necessary to install some additional automated and intelligent devices for the purification, cooling, and circulation of coolants.

3. Low-temperature casting

Low-temperature casting allows molds and models to be cooled by coolants. Generally, liquid nitrogen at −196°C is usually used as a coolant to freeze the water in the sand. The sand particles are frozen together by a film of liquid ice, resulting in a hardened cryogenic mold [10]. This method offers several advantages, including high-quality castings with fine grains, precise dimensions, no smoke or toxic gases during shaping and pouring, and easy sand removal from the mold. Also, it has broad prospects in the fields of nodular cast iron, cast steel, aluminum, copper, and other alloy castings. To illustrate the principles of cryogenic cutting, examples of coolant spraying cutting and coolant immersing cutting are shown in Sections 2.1 and 2.2.

3.1 "Effset" method

This method was invented in 1977 by the foundry of W.H.Booth Company in the United Kingdom. It involves blowing liquid carbon dioxide or liquid nitrogen into the mold to freeze it. After the removal from the box, the mold is placed on an aluminum plate and simultaneously transported through a belt conveyor. It is further fed into a cooling channel to reach the required freezing temperature before assembly. This method gives the mold good collapsibility after being poured, and the molding sand can be easily reused after drying. The principle for freeze casting is shown in **Figure 3**.

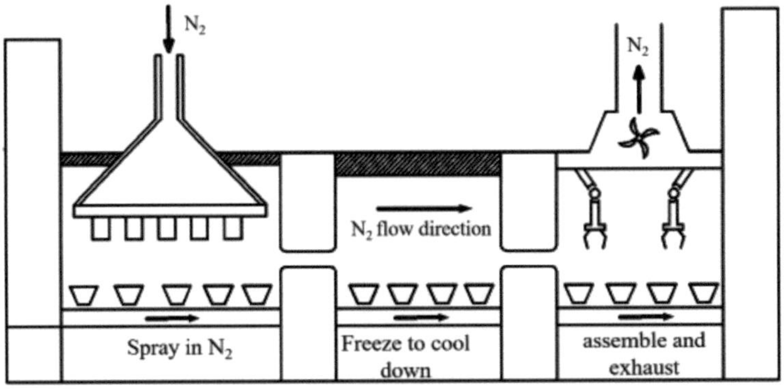

Figure 3.
Principle for freeze casting with a production line [3].

3.2 Cooling tube method

This method is also known as the pressurized LN_2 method. Fine copper tubes with good thermal conductivity are arranged at the top and bottom of the sand mold. Liquid nitrogen in a sealed chamber is then pressurized into the mold at a pressure from 0.5 MPa to 4 MPa, as seen in **Figure 4**.

3.3 Cooling transformation method

This method generally utilizes the LN_2 to cool the intermediate heat carrier, such as Freon. The Freon then freezes the mold and uses a refrigerant to chill the sand mold with a 3–7% moisture [11]. The refrigerant circulates through the evaporator in a cooling system, as seen in **Figure 5**.

Low-temperature casting has many advantages, but there are still some problems: (1) the low production efficiency and extended freezing times; (2) a small amount of binder must be added to the sand to obtain a prestrength of the mold, and hence some

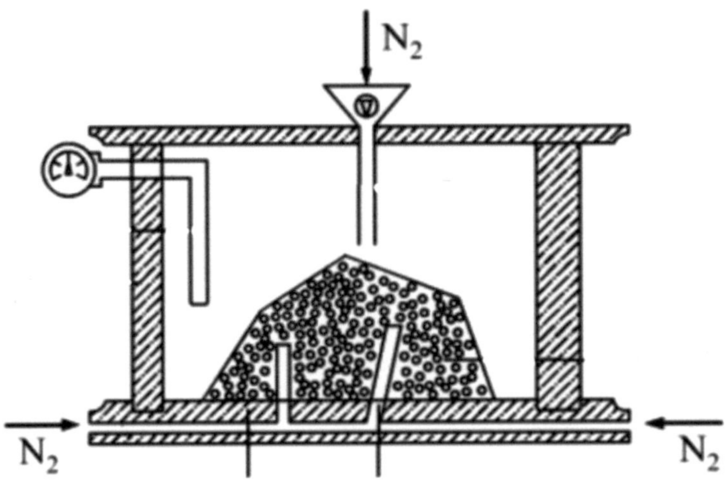

Figure 4.
Principle for the cooling tube method [3].

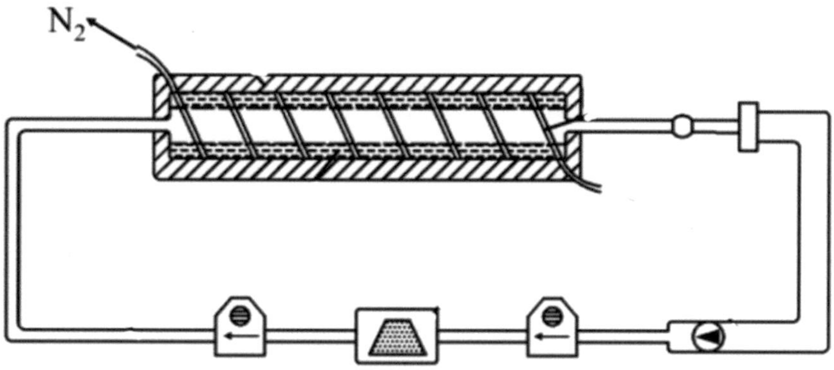

Figure 5.
Principle for the cooling transformation method [3].

advantages of cold molding are lost; (3) the mold is easily deformed by freezing after demolding as the cold mold melts quickly and wastes energy; (4) liquid nitrogen is underutilized with a complicated nitrogen generators and storage containers.

4. Low-temperature treatment

Low-temperature treatment (LTT) involves the use of cooling agents, such as liquid oxygen, liquid nitrogen, and liquid argon. To reduce the volume of residual austenite after quenching and subsequently stabilize the dimensions of the parts, the metal components are immersed in low-temperature coolants within the temperature range from −120 to −196°C. Further, low-temperature treatment can effectively improve the process performance of alloys. For instance, transformation-induced plastics (TRIP) steels are conducted by solution treatment at 1120°C, then plastic deformation at 450°C, and finally cryogenic treatment with LN_2 at −196°C. Its tensile strength can reach 1950 MPa, with an approximate 20% elongation. Its impact toughness and fracture toughness are significantly improved. To a certain extent, low-temperature treatment can improve the size stability of parts, eliminate residual stress, and prevent deformation from the aging [12].

Generally, the sublimation with dry ices plus alcohols is used as a coolant for the low-temperature treatment. In addition, Freon is used for circulating refrigeration in freezers, or liquid oxygen and liquid nitrogen can be evaporated to obtain the low temperature. For most steels, mixtures of dry ice with alcohol and liquid nitrogen come from low costs and wide sources and meet most of the needs of low-temperature treatment. Common coolants and their lowest temperature are shown in **Table 1**.

Coolants	Temperature(°C)	Medium	Temperature(°C)
Dry ice+alcohol	−78	Propane	−42.2
Liquid oxygen	−183	F-12(CF2Cl2)	−29.8
Liquid nitrogen	−195.8	F-13(CF3Cl)	−81.5
Liquid hydrogen	−252.8	F-14(CF4)	−128
Liquid helium	−268.9	F-22(CHF2Cl)	−40.4

Table 1.
Common coolants and their lowest temperature.

Generally, the common temperature range of ordinary LTT is varied from −50 to −80°C. The majority of residual austenite after quenching can be eliminated for simple shape parts, such as bearing, measuring tool, and other tool steels. Further, the hardness and wear resistance of the parts are improved, and the crack is reduced. As shown in **Figure 6**, the hardness of Cr12MoV and CrWMn steels increases significantly after the LTT. Also, it can be used to adjust the out-of-tolerance size for reducing the intersection and cracks. Generally, the volume of treated parts after the LTT can increase due to the transformation of residual austenite into martensite. The optimum of conventional quenching, tempering, and the out-of-tolerance of mold size can be achieved by using this rule. Therefore, LTT had a practical value to improve the service life of tools and improve the quality of parts. However, to greatly avoid cracking defects from temperature stress, the staged LTT is required for the part with complex shapes. The parts with a complex shape, large difference of wall thickness, and small edge distance of holes should be cooled slowly in each stage.

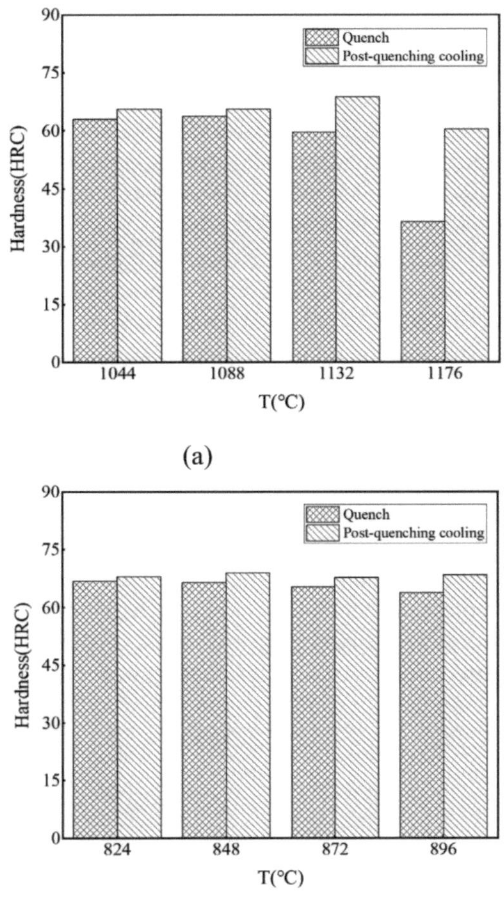

Figure 6.
Hardness of tool steels after quenching and LTT. (a) Cr12MoV (b) CrWMn.

5. Cryogenic forming

5.1 Dual enhancement effect

As shown in **Figures 7** and **8**, the elongation of the annealed sheet at low temperature of −160°C increases to 32.3%, while the solution-treated sheet increases from 24.3 to 34.4%, showing an increase of 41.6% when the temperature is varied from RT to −160°C. Similarly, the strain-hardening exponent for the annealed 2219 aluminum alloy increases from 0.27 to 0.37, showing an increase of 37%. Similarly, the strain-hardening exponent (n) in the solution-treated state enhanced from 0.3 to 0.35 with an increase of 16.7%, when the temperature is varied from RT to −160°C. It is indicated that the sheet shows the dual enhancements of elongation and strain-hardening exponent in low-temperature deformation.

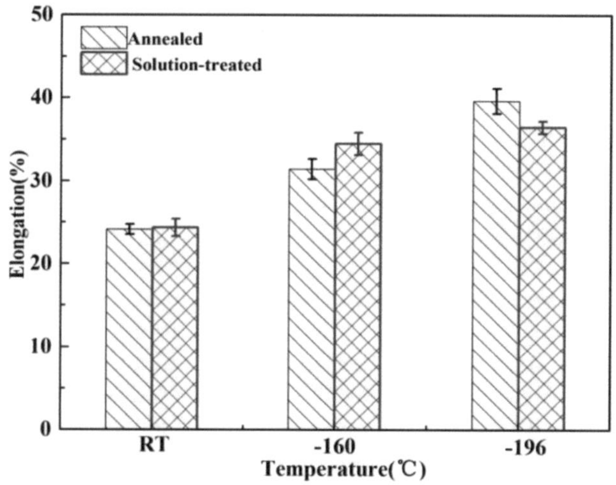

Figure 7.
Elongation of 2219 aluminum alloy varied with low temperatures.

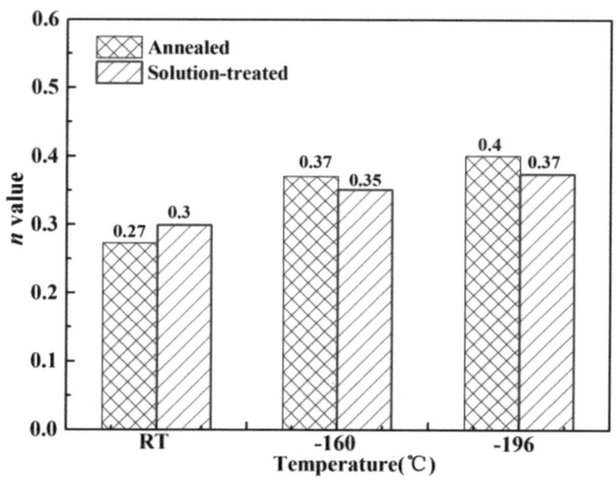

Figure 8.
Strain-hardening exponent (n) varied with low temperatures.

Generally, the strength, hardness, and wear resistance of most metals increase significantly, and their ductility decreases at low temperatures. However, most face-centered cubic (fcc) matrix structure metals, such as Al, Cu, and Ni, do not present cold embrittlement; on the contrary, the elongation and strain-hardening exponent are simultaneously increased during the cryogenic deformation. This abnormal phenomenon is called "dual enhancement effect" [13, 14]. Based on these excellent characteristics mentioned above, problems for the forming of thin-walled aluminum alloy components, such as localized thinning, cracking, and microstructure degradation, can be inhibited. Further, the forming limit and forming process window of the alloys are expanded. Hence, cryogenic forming is transformative and pioneering as compared to the conventional cold and hot formings. It can also lead to a significant increase in the forming potential of fcc metal complex components.

5.2 TRIP effect

Transformation-induced plasticity (TRIP) effect refers to the introduction of phase transformation strengthening and plasticity growth of the stabilized residual austenite in the steel, when it initially transforms to martensite during the deformation. Generally, the martensitic transformation point (Ms) of such steels is lower than room temperature (RT) [15]. As a result, a more pronounced TRIP effect occurs during the cryogenic deformation, as shown in **Figure 9**. The strain is concentrated in defect-rich areas and induces the martensitic nucleation when the austenite stainless steel (ASS) sheet is subjected to certain external loads in low-temperature deformation. Further, the substable residual austenite in steels starts to transform to martensite when aggregated to a certain extent. Since martensite is a hard phase, it increases the local hardness. Continued deformation of the steel becomes difficult. Thereby the deformation is transferred to surrounding zones, and the residual stress is reduced. As the deformation continues, the necking of the material is delayed, and hence the elongation of the material is greatly increased. Additionally, the transformed martensite and increased dislocations during the low-temperature deformation result in a significant increase in the work-hardening capacity of steels.

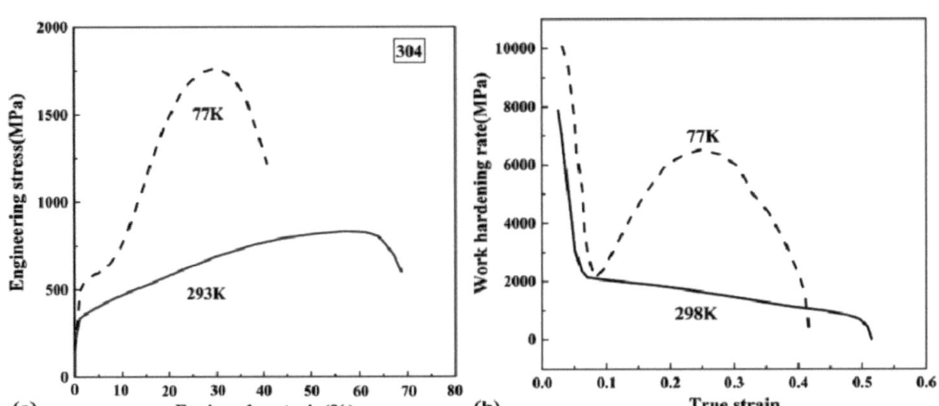

Figure 9.
Tensile curves and work-hardening curves of 304 ASSs.

5.3 Cryogenic forming methods

5.3.1 Bulging

Bulging is a crucial method for the testing of sheet metal formability under a biaxial stress, including hydraulic bulging, rigid punch bulging, electromagnetic bulging, flexible or particle medium bulgings, etc. Currently, hydraulic and rigid die bulgings are the most widely performed, and their corresponding theories are well-developed. However, the bulging at low temperatures is limited for their bulging device and formability of alloy sheets. As shown in **Figures 10** and **11**, the limit for bulging height of 2219 aluminum alloy sheets at −160°C increases by 47.8% as compared to RT [16]. The bulging height and expansion ratio of 6061 aluminum alloy tubes increase nearly double during the cryogenic free bulging [17]. In addition, 304 stainless steel shows a more significant strengthening and more uniform thickness distribution of dome parts at LN_2. Hence, the formability of alloy sheets and part properties can significantly be improved at cryogenic temperatures.

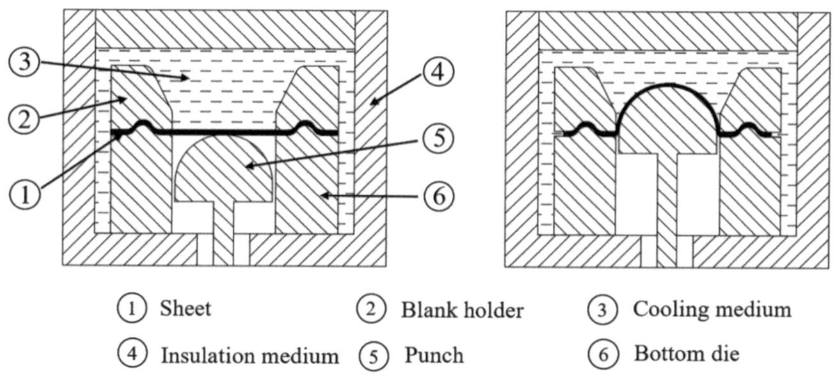

Figure 10.
Schematic diagram of cryogenic bulging of sheets.

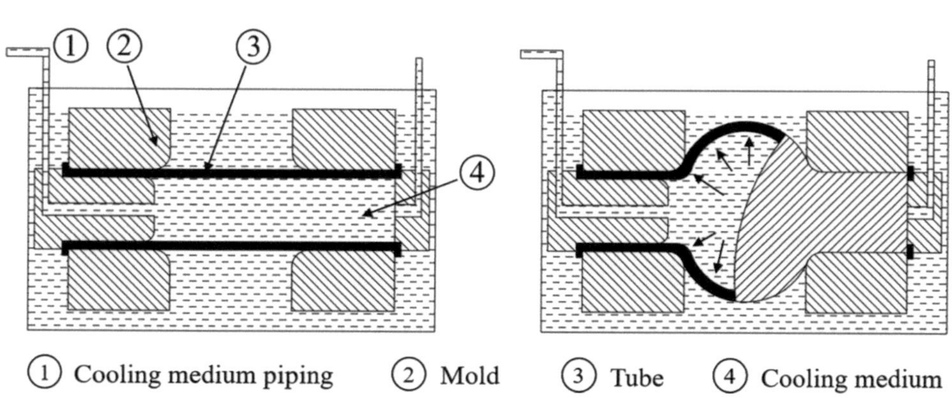

Figure 11.
Schematic diagram of cryogenic bulging of tubes.

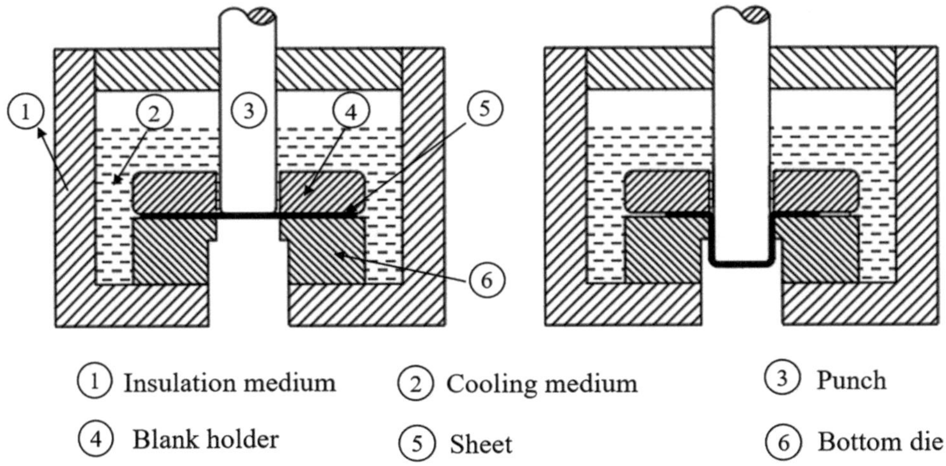

Figure 12.
Schematic diagram of cryogenic deep drawing of cups.

5.3.2 Cup deep drawing

The forming potential of alloy sheets under complex stresses can be evaluated by cup deep drawing, as shown in **Figure 12**. The cups formed at cryogenic temperatures show a greater limiting drawing ratio (LDR). The deep drawing limit of 2219 aluminum alloy increases by 15.6%, and the deep-drawn cups show a relatively more uniform thickness distribution at a low cryogenic temperature of −160°C [18]. An increased resistance to localized plastic flow and thinning occurs for alloy sheets under complex stresses. Cup deep drawing at cryogenic temperatures provides a new approach for the forming of complex thin-walled components.

5.3.3 Bending

Bending is one of the common methods in sheet metal forming. The forming process of complex parts is prone to defects such as springback, wrinkling, and cracking due to the bending stress. This leads to serious constraints on the forming quality and accuracy of complex parts. The most basic forms are V-shaped and U-shaped bendings, as shown in **Figure 13**. The uneven deformation during the bending of thin-walled pipes can lead to excessive springback and instability in tension and compression states. Affecting factors of springback include thickness, elastic modulus, hardening coefficient, die stroke, and strain-hardening exponent. Many bending properties of fcc structured materials are significantly enhanced at cryogenic temperatures. The springback angle becomes greater and bending cracks are suppressed. This imposes higher requirements for the designing of springback compensation to achieve the precise forming.

5.4 Mechanism of cryogenic forming for aluminum alloys

Due to a great stacking fault energy (SFE), the deformation of aluminum alloys is mainly caused by dislocation movement. As shown in **Figure 14**, abundant

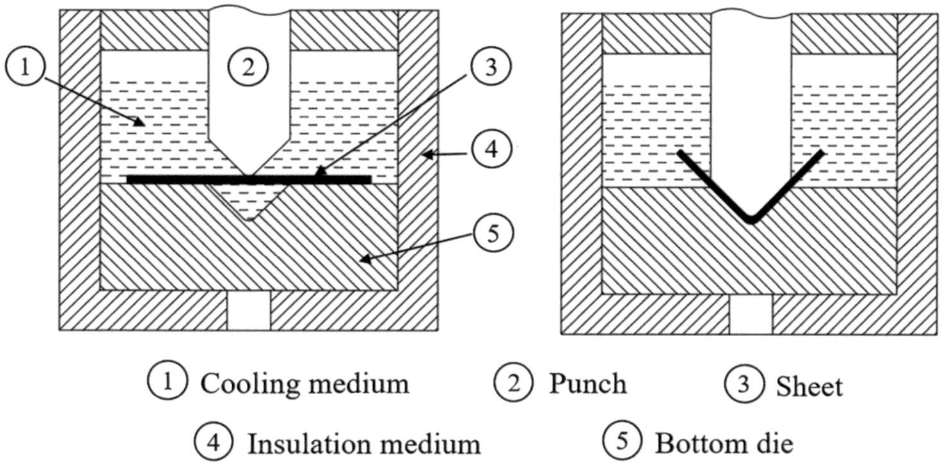

Figure 13.
Schematic diagram of cryogenic bending of sheets.

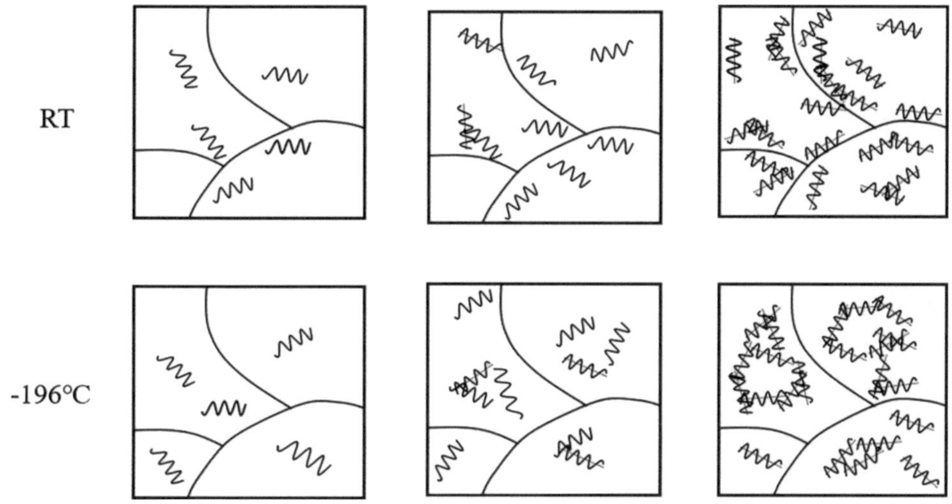

Figure 14.
Microstructure evolution of aluminum alloys at room and cryogenic temperatures.

dislocations are generated in the crystal of aluminum alloys during the cryogenic deformation. Dislocation lines intertwine and polymerize with each other. Likewise, many dislocation networks and cell-like structures are generated. Hence, the cryogenic strengthening effect is markedly increased. In addition, cryogenic temperature can promote uniform dislocation slips and intragranular aggregations. It also leads to the avoidance of crack initiation and localized instability at grain boundaries. The "avalanche effect" of dislocation substructures is suppressed. Hence, the elongation of the cryogenic deformation is increased [19]. According to the Mott hardening model in Eq. (1), the smaller the mean slip distance of dislocations, the more likely the activation of a large number of dislocation substructures. In return, the formation of dislocation cells and dislocation networks can be promoted. Macroscopic strain localization can be hindered. Hence, the strain-hardening capacity of these alloys is enhanced.

$$\chi_i = \frac{G^2 b}{2\pi^2 \lambda_i} \quad (1)$$

where χ_i is the strain-hardening coefficient, b is the Burgers vector, λ_i is the mean slip distance, and G is the shear modulus.

5.5 Mechanism of cryogenic forming for austenitic steels

The martensitic transition is a nondiffusive transition that occurs in austenitic steels during cryogenic deformations. Martensite formation is realized by a shear mode, and the atoms at the interface between martensite and austenite are shared and movable. The entire interface is maintained coherent by parent-phase shear mode [20]. In addition, a series of substructures, such as dislocations, twins, and stacking fault, can be induced by localized plastic deformation. Martensite transformation is caused by chemical driving forces when the temperature is below the onset of the phase transformation (Ms) [15]. As seen in **Figure 15**, the mechanical driving force for the martensitic transformation is provided by the deformation energy. The interfacial and strain energy of the new phase are provided by their superposition during the cryogenic forming. It further results in greatly enhanced properties of austenitic steels due to the high hardness and strength of the martensitic phase. Previous research by the author found that the cooperative strengthening by dislocations and α′-martensitic transformation are more dominant during the LN_2 deformation. As seen in **Figure 16**, the undulated work-hardening behavior

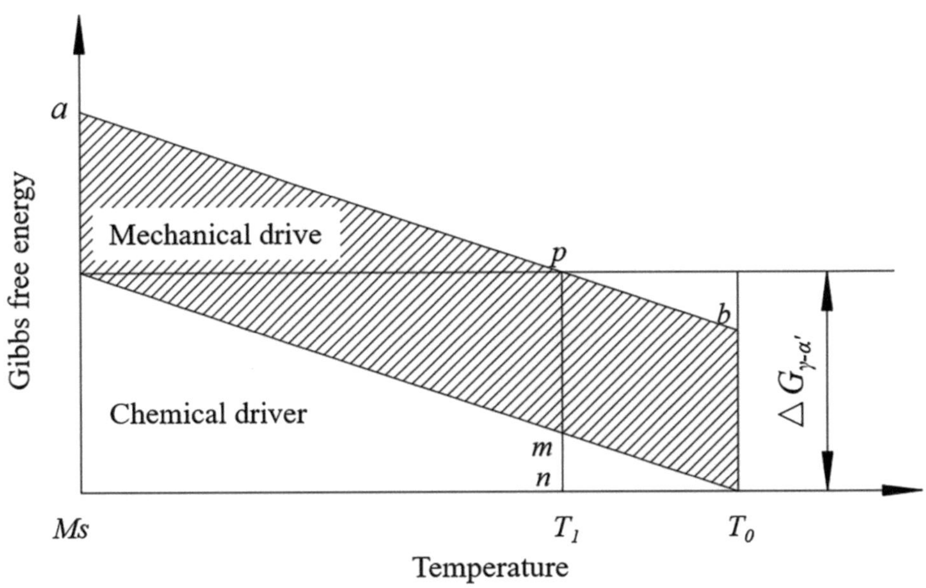

Figure 15.
Schematic diagram of thermodynamic conditions for martensitic transformation.

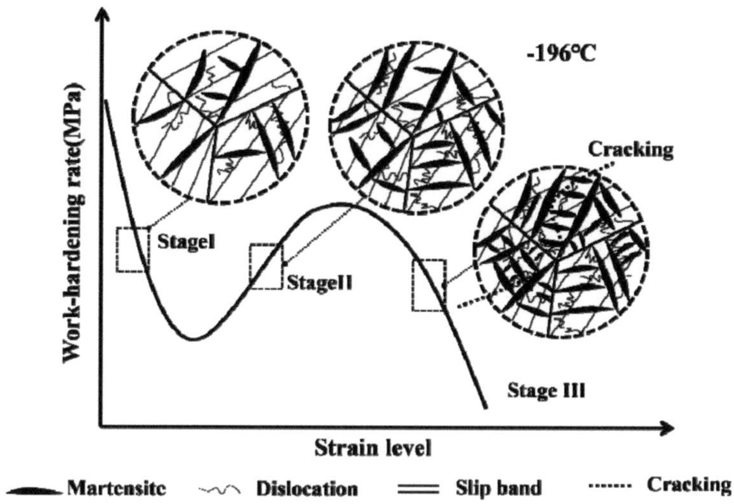

Figure 16.
Microstructure evolution of ASS sheets at cryogenic temperatures [15].

occurs due to the preferentially promoted α′-martensitic phases and decreased austenite phases. Likewise, the distribution of martensite phase is inhomogeneous, accompanied by localized dense dislocations. Hence, the austenitic steels are exhibited by a low elongation and undulated work-hardening during the LN_2 deformation.

6. Cryogenic forming applications

Cryogenic forming of alloy sheets is a manufacturing process that is carried out below room temperature or very low temperatures. Generally, the alloy sheets should be precooled to an ultra-low temperature range for −150°C to −196°C via a coolant. And then the deep-cooled sheet is subjected to plastic deformation through a die device in order to form complex-shaped components. Compared to conventional hot forming processes, cryogenic forming has the following advantages: (1) energy saving and environment friendly. Conventional hot forming requires a high temperature or high pressure, while the cryogenic forming only requires coolants for achieving temperature conditions. This method does not consume electricity or avoids industrial waste gas; (2) high product quality and production efficiency. The superior cooling effect can be quickly achieved by adjusting the flow rate of the coolant. Therefore, it can achieve fast process efficiency and reduce production cycles; (3) wide potential for applications. Cryogenic forming can be used for various alloys and presents a good universality, and low cost for manufacturing industries, such as aerospace components, automotive parts, electronic product casings, and furniture.

As shown in **Figure 17**, localized thinning, cracking, and microstructure degradation of aerospace aluminum alloy components can be inhibited, and their forming limit and process window can be expanded. Automotive parts usually need complex processes, and low-temperature forming can make these complex

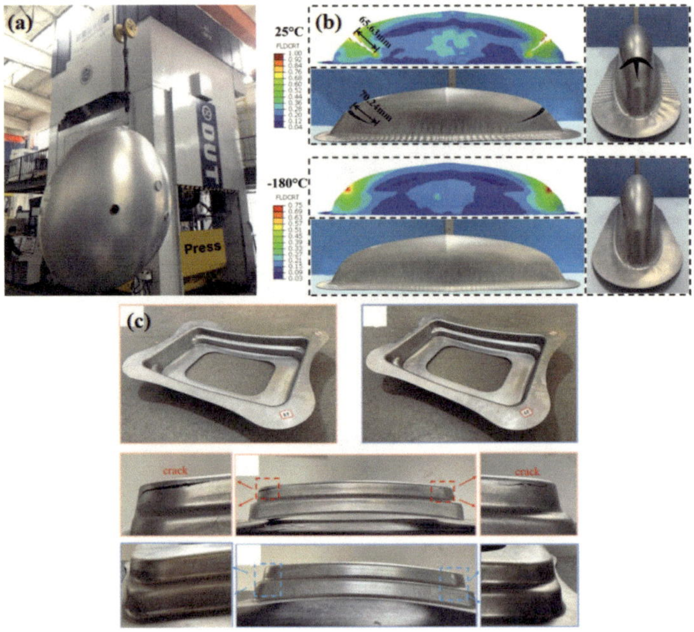

Figure 17.
Typical aluminum alloy components via cryogenic formings: (a) Rocket tank [14]; (b) thin-walled components [21]; (c) hollow box-shaped components [22].

components easier to form and greatly improve production efficiency. Electronic product casings via cryogenic forming can produce exquisite casings with a significant cryogenic lubrication effect and further ensure the performance and quality of electronic products.

7. Conclusions

Low-temperature manufacture is a new revolution in manufacturing industries, which can help reduce energy consumption, improve production efficiency and product quality, and minimize environmental impact. With the advancement and innovation of this technology, low-temperature manufacture will become increasingly mature, and their applications will become more and more extensive. The main conclusions are as follows:

1. Low-temperature cutting is a green cutting method for reducing the cutting temperature, delaying the tool wear, reducing the surface roughness, and increasing the surface hardness and residual compressive stress. However, the current applications for this method are still deficient.

2. High-quality castings with fine grains, precise dimensions, and no smoke or toxic gases can be achieved via low-temperature casting. However, applications of low-temperature casting is limited for its low production efficiency and extended freezing times. There is still a lack of research on the mechanism of low-temperature casting.

3. Through the low-temperature treatment of steels, such as liquid nitrogen from −120 to −196°C, the residual austenite of metals can be reduced, and the hardness, wear resistance, and stability can be improved.

4. Most fcc metals can be significantly strengthened at cryogenic temperatures. Further, the elongation and strain-hardening exponent at cryogenic temperatures increase significantly, especially for aluminum alloys. Significant changes in work hardening behaviors can also be caused by the cryogenic forming. In the future, the manufacturing industry will increasingly focus on energy conservation, environmental protection, and efficiency improvement. And hence low-temperature technology will become a preferred choice for various manufacturing industries.

Author details

Wangjun Cheng
School of Mechanical Engineering, Xinjiang University, Urumqi, People's Republic of China

*Address all correspondence to: chengwangjun@xju.edu.cn; chengwangjun2008@126.com

References

[1] Blagg K, Castagnede A, Singh M. On-chip heating effects in electronic measurements at cryogenic temperatures. Cryogenics. 2022;**126**:103536

[2] Yuan SJ. Fundamentals and processes of fluid pressure forming technology for complex thin-walled components. Engineering. 2021;**07**:358-366

[3] Li WB. Low Temperature Application Engineering. Beijing: Weapon Industry Press; 1992

[4] Cheng WJ, Gao Q, Sun YN, et al. Research progress of freezing processes and devices for fresh meat products. International Journal of Refrigeration. 2024;**161**:71-82

[5] Alcaide F, Sirés I, Brillas E, et al. Coupling wastewater treatment with fuel cells and hydrogen technology. Current Opinion in Electrochemistry. 2024;**45**:101530

[6] Gupta MK, Niesłony P, Korkmaz ME, et al. Potential use of cryogenic cooling for improving the tribological and tool wear characteristics while machining aluminum alloys. Tribology International. 2023;**183**:108434

[7] Yildiz Y, Nalbant M. A review of cryogenic cooling in machining processes. International Journal of Machine Tools & Manufacture. 2008;**48**:947-964

[8] Wu BH, Cui D, He XD, et al. Cutting tool temperature prediction method using analytical model for end milling. Chinese Journal of Aeronautics. 2016;**29**(6):1788-1794

[9] Venugopal KA, Paul S, Chattopadhyay AB. Tool wear in cryogenic turning of Ti-6Al-4V alloy. Cryogenics. 2007;**47**:12-18

[10] Minowa S, Ninomiya M, Ohta H. Deformation and pattern drawing force of sand mold at freezing. Journal of Japan Foundry Engineering Society. 1981;**53**:15-19

[11] Shao G, Hanaor DAH, Shen X, et al. Freeze casting: From low-dimensional building blocks to aligned porous structures-a review of novel materials, methods, and applications. Advanced Materials. 2020;**32**(17):1907176

[12] Litao C, Kudzanai M, Jonathan D, et al. The effect of low temperature heat treatment on stress corrosion crack initiation in machined 316L stainless steel in high-temperature hydrogenated water. Scripta Materialia. 2021;**195**:113742

[13] Cheng WJ, Liu W, Fan XB, et al. Cooperative enhancements in ductility and strain hardening of a solution-treated Al-Cu-Mn alloy at cryogenic temperatures. Materials Science and Engineering: A. 2020;**790**:139707

[14] Fan XB, Chen XS, Yuan SJ. Novel forming process for aluminum alloy thin shells at ultra-low temperature gradient. International Journal of Machine Tools and Manufacture. 2023;**185**:103992

[15] Cheng WJ, Cui DD, Sun YN, et al. Cryogenic work-hardening behavior for a metastable austenitic stainless steel at liquid nitrogen temperature. Materials Science and Engineering: A. 2022;**861**:144352

[16] Cheng WJ, Liu W, Yuan SJ. Deformation behavior of Al-Cu-Mn alloy sheets under biaxial stress at cryogenic

temperatures. Materials Science and Engineering: A. 2019;**759**:357-367

[17] Wang XG, Fan XB, Chen XS, Yuan SJ. Cryogenic deformation behavior of 6061 aluminum alloy tube under biaxial tension condition. Journal of Materials Processing Technology. 2022;**303**:117532

[18] Yuan SJ, Cheng WJ, Liu W, Xu YC. A novel deep drawing process for aluminum alloy sheets at cryogenic temperatures. Journal of Material Processing Technology. 2020;**284**:116743

[19] Liu W, Cheng WJ, Yuan SJ. Analyses on formability and flow stress of an Al-Cu-Mn alloy sheet under biaxial stress at cryogenic temperatures. International Journal of Mechanical Sciences. 2021;**195**(5):106266

[20] Shen YF, Li XX, Sun X, et al. Twinning and martensite in a 304 austenitic stainless steel. Materials Science and Engineering: A. 2012;**552**:514-522

[21] Wang CG, Yi YP, Wang HH, et al. Investigation on the formability and deformation mechanism of aluminum alloy thin-walled parts at cryogenic temperature. Journal of Materials Processing Technology. 2023;**319**:118041

[22] Dong F, Huang SQ, Yi YP, et al. Enhanced deep drawing formability and deformation mechanism of aluminum alloy at cryogenic temperature. Journal of Alloys and Compounds. 2023;**968**:171992